Molecular plant development

Molecular plant development

from gene to plant

by

Peter Westhoff

Holger Jeske, Gerd Jürgens,
Klaus Kloppstech, Gerhard Link

With illustrations by
Melanie Waigand-Brauner, Freiburg, Germany

Translated by
Ellen Peerenboom, Köln, Germany
and Ewen Cartwright, Harpenden, UK

OXFORD • NEW YORK • TOKYO
OXFORD UNIVERSITY PRESS
1998

Oxford University Press, Great Clarendon Street, Oxford OX2 6DP

Oxford   New York
Athens   Auckland   Bangkok   Bogota   Bombay
Buenos Aires   Calcutta   Cape Town   Dar es Salaam
Delhi   Florence   Hong Kong   Istanbul   Karachi
Kuala Lumpur   Madras   Madrid   Melbourne
Mexico City   Nairobi   Paris   Singapore
Taipei   Tokyo   Toronto   Warsaw
and associated companies in
Berlin   Ibadan

Oxford is a trade mark of Oxford University Press

Published in the United States
by Oxford University Press Inc., New York

First published in 1996 by Georg Thieme Verlag under the title Molekulare Entwicklungsbiologie.

© Georg Thieme Verlag, Stuttgart, 1998

A catalogue record for this book is available from the British Library

Library of Congress Cataloging in Publication Data
(Data applied for)

ISBN 0 19 850204 4 (Hbk)
ISBN 0 19 850203 6 (Pbk)

Typeset by EXPO Holdings, Malaysia

Printed in Great Britain by The Bath Press, Avon.

Preface to the English edition

Almost two years after the publication of the German edition of this textbook it is rewarding to note that it has been well received by both teachers and students in plant biology. There must have been an obvious demand for a text which deals with this modern and topical aspect of botany. When we were approached and given the option to assist with an English version of the book we realized that it required more than just a translation for the Anglo-American student. Taking account of this, much of the text has been re-written for the English edition, and many of the, much appreciated, comments from colleagues and students have been acted upon. We hope that these changes and the inclusion of recent developments will benefit the reader. The literature has been surveyed and updated to include results published by November 1997. During the final revision some additional references to primary literature dated January 1998 have been added. This should make this English version of the book not only a very comprehensive text on plant development but also one which is up to the minute on the current state of the art in this rapidly expanding field of the plant sciences.

Thanks are due to the translators who had the difficult task to convert the long German sentences with many sub-clauses into readable and, hopefully, comprehensible English. The authors would greatly appreciate any suggestions for improvement from the readers of the English text and hope that they provide as many helpful comments as their German colleagues. Dr U. J. Santore from my department has been kind enough to clarify many of my linguistic queries and revised a couple of passages in the text.

Düsseldorf   P. W.
January 1998

Preface

'Can you recommend us a textbook that deals with the subject matter of your lectures on molecular and developmental biology of plants?' For years students have asked me this question and received only an apologetic head-shake in response. When a publisher also points out the lack of such a textbook, then taking action becomes almost obligatory. To this end, five university lecturers competent in different areas of research in this field have collaborated in writing a textbook on molecular developmental biology. In this way the task was shared among several people, ensuring an accurate reflection of the current state of knowledge in such a rapidly developing field of research.

Books with joint authorship run the risk of not presenting a whole and unified picture of their subject, as each author's individual way of seeing and presenting his or her topic can be too strongly accentuated. We have dealt with this difficulty by having a clearly defined concept of content and form from the very outset. Moreover, each contribution was harmonized with the others after completion. The reader can judge for him/herself how successful we have been in this respect. As biologists, we are aware that feedback is the key to the successful optimization of a system and we would therefore like and positively encourage you, the users of this book, to let us have your criticisms.

The reader will find our concept of the structure of the book's contents described in the introductory chapter.

In order to accomplish the objectives we set ourselves, each chapter is preceded by a simple outline that permits quick reference for key questions and themes. Each chapter is structurally divided into a preface, basic knowledge, definitions, and supplementary information.

- The *preface* of each chapter serves as a compass for orientation. Here the subject is outlined and its structure explained.
- *Basic knowledge* is provided in the text that follows. The corresponding illustrations and tables aid in assimilating the subject matter.
- *Definitions* of important key concepts are explained in the form of brief, visually highlighted descriptions. Specialized vocabulary is provided in a glossary at the back of the book.

- *Supplementary information* is provided in boxes. These permit a more intensive treatment of the information covered and, in addition, provide information about current research status, methods, and historical developments.

Finally we would like to warmly thank those whose generous help has greatly facilitated our work on the manuscript: we apologize to our families and colleagues for our long neglect—important things have been left unattended to and we shall now hopefully catch up on them! We thank our colleagues for their frequent advice, and our secretaries for their valuable help with typing. In particular, I wish to thank Marianne Limpert, who assisted in compiling the 'very last draft' of the entire manuscript with great understanding and good humour.

Düsseldorf   P. W.
November 1996

Contents

List of authors *xi*

1 A brief introduction to this textbook *1*
P. Westhoff
1.1 What are the aims of this textbook and who is it intended for? *1*
1.2 How is this textbook organized? *2*
1.3 How do I use this textbook and how do I obtain further information? *3*

2 What is development? *6*
G. Link
2.1 Some background *7*
2.2 From simple to complex development *9*
2.3 Key processes of development: growth, differentiation, and morphogenesis *13*
2.4 Reasons for cell differentiation *20*
2.5 Regulation of development in multicellular organisms *25*
2.6 Development in plants and animals differs because plants possess meristems *34*
2.7 Bibliography *37*

3 The genetic analysis of developmental processes — methodology *39*
G. Jürgens
3.1 Developmental genes can be recognized by their mutant phenotypes *42*
3.2 From phenotype to gene — how developmental genes can be isolated *52*
3.3 From the gene back to the phenotype — molecular analyses of the effects of developmental genes *62*
3.4 Prospects *64*
3.5 Bibliography *65*

4 The genetic material of plant cells is compartmentalized: structure and expression of the subgenomes *67*
P. Westhoff
4.1 The nuclear genome *70*
4.2 The plastome *87*
4.3 The chondriome *101*
4.4 Interactions between different genetic compartments *104*
4.5 Bibliography *107*

5 **Light, phytohormones and the biological clock as inducers and modulators of development** *110*
P. Westhoff and K. Kloppstech

5.1 What is light and how does it affect the plant's life cycle? *111*

5.2 Phytochrome as a prototype for plant photoreceptors *112*

5.3 Blue light—a system that induces developmental processes and protects against light stress *127*

5.4 The biological clock *130*

5.5 Phytohormones *133*

5.6 From the seed to the young seedling: phytohormones and light as regulators of development *145*

5.7 Bibliography *157*

6 **Phases during the life cycle of the flowering plant** *161*
G. Jürgens

6.1 Embryogenesis *163*
G. Jürgens

6.2 Postembryonic vegetative development *176*
G. Link

6.3 The generative phase *191*

6.4 Bibliography *215*

7 **Pathogens and symbionts as growth modulators** *220*
H. Jeske

7.1 Viruses and viroids *222*

7.2 Agrobacteria *235*

7.3 Rhizobia *244*

7.4 Bibliography *253*

Glossary *256*

Index *263*

Authors

H. Jeske
Biologisches Institut, Abteilung Botanik, Pfaffenwaldring 67, 70550 Stuttgart, Germany

G. Jürgens
Universität Tübingen, Lehrstuhl für Entwicklungsgenetik, Spemannstrasse 37–39, 72076 Tübingen, Germany

K. Kloppstech
Institut für Botanik der Universität Hannover, Herrenhäuser Strasse 2, 30419 Hannover, Germany

G. Link
Lehrstuhl für Pflanzenphysiologie, Ruhr-Universität, Universitätsstrasse 150, 44801 Bochum, Germany

P. Westhoff
Institut für Entwicklungs- und Molekularbiologie der Pflanzen, Heinrich-Heine-Universität, Universitätsstrasse 1, 40225 Düsseldorf, Germany (email: west@uni-duesseldorf. de)

1 A brief introduction to this textbook

Contents

1.1 What are the aims of this textbook and who is it intended for?

1.2 How is this textbook organized?

1.3 How do I use this textbook and how do I obtain further information?

1.1 What are the aims of this textbook and who is it intended for?

Until the early 1980s the study of plant development was essentially restricted to a morphological and physiological description of developmental processes. Only with the emergence of molecular biology, and particularly with the implementation of genetic methods, has modern developmental plant biology been orientated towards the gene. Now questions arise as to which genes play a role in developmental processes and how they work at the molecular level. Just how successful this genetic–molecular biological approach has been in the analysis of developmental processes in animals is made evident by the impressive experiments that have been carried out on *Drosophila melanogaster* and *Caenorhabditis elegans*. Only with this approach has it become possible to progress from descriptive developmental biology to causal analysis. The present textbook therefore concentrates upon this up-to-date and potentially fruitful approach to developmental analysis in plants.

Of course, a genetic–molecular biologically orientated developmental plant biology makes use of the numerous accounts of developmental processes in the most varied plant species as they have been compiled during past decades. These findings still constitute an important source for the selection of suitable experimental systems. However, nowadays the tendency is to concentrate upon a few model plants such as *Arabidopsis thaliana* or *Zea mays*. In this way, it is hoped that fundamental and universally valid discoveries about developmental processes in plants will be reached more rapidly than they would be by directing efforts towards many different systems. The experiments hitherto undertaken on model systems for animals support these expectations; indeed, they show that fundamental underlying developmental processes have remained relatively unaltered throughout evolution.

Textbooks are always expected to be all-encompassing. Of course, they should present the widely accepted fundamental principles of their subject. However, if they are limited in this way, they easily lose the readers' interest, as the impression is given that everything of essential interest has already been researched. For this reason, textbooks should always provide scope for some insight into the current state of knowledge in their field, especially when the subject of the textbook—in this case 'developmental plant biology'—is undergoing a turbulent period of growth. This textbook therefore consciously seeks to include recent research developments in the field. Consequently, several of the statements are as yet provisional and must be supplemented or modified at a later stage. In this way, it is made apparent that molecular developmental plant biology is still in flux and students will gain an idea of the dynamics of this field of research.

When we speak of 'students', we do not really mean the novice student who has just arrived at university. We address ourselves far more to the advanced student who is already familiar with the fundamentals of general botany and plant physiology and who possesses, moreover, a sound basic knowledge of molecular biology and genetics. If these are lacking, then they should be acquired through the (parallel) study of appropriate textbooks. Only these conditions have made it possible to limit the scope of this textbook to manageable proportions.

1.2 How is this textbook organized?

The textbook begins with a basic introduction to the question of plant development. Here the framework is outlined and important concepts and findings of classical plant biology are presented. As this book lays emphasis upon the genetic–molecular biological strategy for the analysis of developmental processes, the reader is then immediately introduced to the methodology of this procedure: how can a genetic approach be used to analyse a developmental process and eventually isolate the participating genes? The isolation and characterization of genes naturally assumes prior knowledge about the plant genome and its expression. The chapter that follows aims to provide this information. The reader will learn that the genetic matter of a plant cell is divided into three different subgenomes whose expression must be co-ordinated during development. At the same time, the evolution of the genome and its expressive mechanisms are also examined.

In the next two chapters the principal developmental processes in the life of the flowering plant are dealt with. Here it becomes evident that development is not merely the result of a rigid genetically determined programme, but that external and internal factors may intervene as inductors and modulators. Light should be mentioned as the most significant external factor, which acts as a signal both in the early phase of shoot development and also later in the transition from the vegetative to the generative phase. Plant hormones, on the other hand, represent the 'classical' endogenous factors, the effects of which we are only just beginning to understand with the application of genetic and molecular biological methods.

The development of a flowering plant is not only influenced by abiotic external factors, but also by interaction with other living organisms, particularly micro-organisms. This can lead to mutually useful co-operation for both parties, as exemplified by rhizobium–legume symbiosis. Often, though, the relationship is rather of unilateral use to the micro-organism and results in pathological developmental processes in the plant. A good example of this is the soil bacterium *Agrobacterium tumefaciens*. The final chapter aims to examine such developmental strategies which are derived from interaction with other living organisms. This also demonstrates that although development is primarily an individual achievement, plant, like all living organisms, form part of a larger network of activity.

1.3 How do I use this textbook and how do I obtain further information?

A textbook that attempts to present a rapidly changing field of knowledge in concentrated form runs the risk of not doing justice to the complexity of the subject and of conveying a crude outline of the current state of research. In order to avoid this, we have added an annotated bibliography at the end of each chapter, which may prompt deeper investigation into a subject.

Primary literature is understood as being the original publications in which scientists communicate the results of their research to those specialized in the field. Table **1.1** sets out a selection of journals in which publications in the field of developmental and molecular plant biology regularly appear. Original publications are often too specialized for the beginner and understanding them demands a certain familiarity with the field of research in question. Secondary literature, on the other hand, is certainly recommended as worthwhile reading even for the beginner. In a review, a summary of the original publications pertaining to a particular topic can be found and a quick overview of a subject area can be gained. More extensive reviews that deal with topics exhaustively and also provide a complete review of the literature can be found in the annually published volumes of the 'Annual Review' series (Table **1.1**). Shorter review articles can be found in specialized, mainly monthly publications, such as 'Trends in' (Table **1.1**), but these can now also be found in those series that were originally reserved for original publications (see journals in Table **1.1**). Intensive and regular study of such review articles cannot be encouraged too early in students.

Reviews are a good starting point for familiarizing oneself with a new topic. However, they are often insufficient if an exhaustive review of literature is required, which also includes the most recent publications. For this purpose database searches are available which can be consulted on-line from a desktop PC and also from university libraries. The results of such database searches can be transferred to one's own personal database. Database systems for the administration of bibliographical references are supplied commercially by several manufacturers (for example, Reference Manager™). Such bibliographical databases make it possible to draw up bibliographies of any edition format and are a great help when writing dissertations, theses, or publications.

In order to keep abreast of development's in one's subject area, one should read regularly the contents summaries of the largest journals. These contents summaries are available in electronic form (for example, Current Contents™ or Reference Update™) and can be consulted according to different search criteria. The search results can be transferred directly into bibliographical databases, so one can build up a well-stocked archive of bibliographical references for one's own field of interest.

As mentioned above, the authors of this textbook believe that molecular and developmental plant biology is an exciting area of research. We therefore hope that this textbook will not only be a useful tool in the preparation for tests and examinations, but that it will prompt the reader to look further and more deeply into this fascinating area of modern plant science.

General journals for original publications

Cell
Development
Developmental Biology
EMBO Journal
European Journal of Biochemistry
Genes and Development
Journal of Biological Chemistry
Molecular and General Genetics
Nature
Nucleic-Acids Research
Proceedings of the National Academy of Sciences (USA)
Science

Specialized plant journals for original publications

Planta
Plant Journal
Plant Molecular Biology
Plant Physiology
The Plant Cell

More extensive review articles

Annual Review of Biochemistry
Annual Review of Cell Biology
Annual Review of Genetics
Annual Review of Phytopathology
Annual Review of Plant Physiology and Plant Molecular Biology

Short review articles

Current Opinion in Genetics and Development
Current Opinion in Plant Biology
Trends in Biochemical Sciences
Trends in Cell Biology
Trends in Genetics
Trends in Plant Science

Table 1.1 Selected journals in which original publications and review articles in the subject area of molecular and developmental biology appear on a regular basis. The journals are presented in alphabetical order in each of the lists.

2 What is development?

Contents

2.1 **Some background**

2.2 **From simple to complex development**
2.2.1 Life cycles and alternation of generations
2.2.2 Simple life cycles
2.2.3 The life cycle of higher land plants

2.3 **Key processes of development: growth, differentiation, and morphogenesis**
2.3.1 Growth: the increase of size and weight
2.3.2 Differentiation: specialization of form and function
2.3.3 Morphogenesis: developing the form of the complete organism

2.4 **Reasons for cell differentiation**
2.4.1 Chemical, electrical, or mechanical signals?
2.4.2 Genes and gene expression
2.4.3 Differential gene expression: from nucleotide sequence information to time- and space-dependent cell activities
2.4.4 Cell polarity

2.5 **Regulation of development in multicellular organisms**
2.5.1 Determination: the commitment to differentiation fates
2.5.2 Pattern formation
2.5.3 Cell–cell contact
2.5.4 Positional information and its interpretation
2.5.5 Long-distance effects

2.6 **Development in plants and animals differs because plants possess meristems**
2.6.1 Embryos
2.6.2 Meristems: the basic units of the plant's developmental plasticity
2.6.3 Cell lineages

2.7 **Bibliography**

Preface

'Development' is such a familiar term it seems unnecessary to spend time explaining it. However, questions and difficulties arise with a more in-depth examination of the word. As we will see, the change in form and function that accompanies development is one of the basic features of biological systems. At the same time the term can be used in a figurative sense, for example, when we talk about the 'development' of a town. Obviously, there are substantial differences in the ways that different groups of organisms change their shape. Also, not every change of form or function in an individual is necessarily a part of its development.

Because the definition of 'development' is somewhat hazy, we will explain it step by step. To help us, the first part of the chapter will discuss some important basic ideas in other areas of biology. To cover these particular areas completely is beyond the scope of this text. We will go on to examine classical developmental biology, and if at all possible, come up with a more satisfactory definition of development. Deliberately, this chapter has not been limited to the development of plants, but will describe common aspects of development for many different organisms. The conclusions will be useful later when we consider modern developmental biology, as it is biased more towards molecular biology and genetics.

2.1 Some background

We are all aware of the basic characteristics of plants and animals. However, there are more than two million species and a countless variety of different shapes. Not all of these shapes will fit into the preconceived notions about an animal shape or a plant shape. This has been problematic for a long time. There have been many attempts at an optimal classification of all organisms, based on comparative biological investigations. These studies have led to the recognition of three major lineages in the tree of life: eubacteria, archaebacteria, and eukaryotes. Each of these lineages may be further subdivided. For instance, the eukaryotes diverged into several complex assemblages: plants, animals, fungi, and the protists, which form a phylogenetically heterogeneous group.

Eubacteria and archaebacteria are composed of a single compartment which is bordered by a membrane and a cell wall. The genetic material of these cells, a DNA–protein complex called a *nucleoid*, is not separated from the rest of the cell by a membrane but is in direct contact with the metabolic processes in the cell. Since eubacteria and archaebacteria do not compartmentalize their genome they are grouped together as prokaryotes.

In contrast, the eukaryotic cell is divided into distinct compartments and so is its genetic material. The majority of DNA is located in a specialized organelle, the nucleus, which is bounded by a double membrane (Figure **2.1**). The nucleus contains species- and cell-specific numbers of DNA–protein complexes called *chromosomes*. Two other organelles, plastids (found in land plants and algae) and mitochondria also contain DNA, albeit in much smaller amounts than in the nucleus. DNA in these organelles is complexed with proteins, but the complexes resemble bacterial nucleoids more than eukaryotic chromosomes.

The prokaryotic-like organization of the genetic material in plastids and mitochondria can be explained by their evolutionary origin. There is enough convincing evidence that plastids and mitochondria originated from ancient, once free-living eubacteria. These were taken up by a eukaryotic host, eventually leading to the evolution of a photosynthetic, aerobic eukaryote.

Besides these two primary endosymbiotic events, several secondary endosymbioses occurred during the evolution of the various lineages of photosynthetic eukaryotes. Cryptophyte algae, for instance, are derived from the

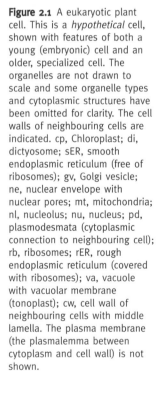

Figure 2.1 A eukaryotic plant cell. This is a *hypothetical* cell, shown with features of both a young (embryonic) cell and an older, specialized cell. The organelles are not drawn to scale and some organelle types and cytoplasmic structures have been omitted for clarity. The cell walls of neighbouring cells are indicated. cp, Chloroplast; di, dictyosome; sER, smooth endoplasmic reticulum (free of ribosomes); gv, Golgi vesicle; ne, nuclear envelope with nuclear pores; mt, mitochondria; nl, nucleolus; nu, nucleus; pd, plasmodesmata (cytoplasmic connection to neighbouring cell); rb, ribosomes; rER, rough endoplasmic reticulum (covered with ribosomes); va, vacuole with vacuolar membrane (tonoplast); cw, cell wall of neighbouring cells with middle lamella. The plasma membrane (the plasmalemma between cytoplasm and cell wall) is not shown.

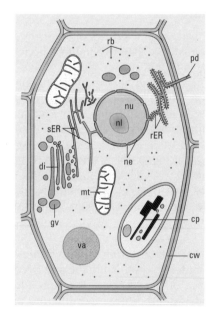

endosymbiotic uptake of a rhodophytic alga by a heterotrophic eukaryotic host. The remnants of the nuclear genome of the rhodophyte, called the *nucleomorph*, are still found in the cryptophytic cell and they are transcribed.

Plants possess basic attributes that are common to many forms of life. These are:

(1) metabolism and transfer of energy;
(2) storage and transmission of genetic information;
(3) cellular structure and intracellular compartmentalization; and
(4) development—each individual will go through certain changes of form and function during its life.

Developmental changes are governed by time and the surrounding environment; although there may be a degree of flexibility in the nature of such changes, they will always be confined to certain limits and depend on the plant species.

Definition Development is defined as the total sum of species-dependent changes of form and function during an individual's life cycle.

It is questionable whether this reductionist definition is adequate if we wish to describe all possible cases of development. If we consider the many different organism forms, as well as the entirely different species-specific changes of form during individual life cycles, the above definition is definitely insufficient. For this reason, it will be useful to consider just the differences in the development of individual organisms (*ontogeny*).

2.2 From simple to complex development

2.2.1 Life cycles and alternation of generations

Ontogeny of individuals does not always lead directly from a starting point A to an end-point B—usually there are several steps in between. Development does not stop once adulthood has been reached. For one thing, senescence begins gradually and limits the individual's activities until finally death occurs. More importantly, adults begin reproducing to maintain their species. From this perspective, development is not a one-way street but is cyclic, following the reproductive cycles of the species.

In single-celled organisms the cell is responsible for the organism's reproduction. During the life cycle of multicellular organisms, special reproductive cells/tissues/organs develop. Theoretically, there are two kinds of reproduction: *asexual* (vegetative) and *sexual* (generative). In the case of asexual reproduction, the propagative units originate from mitosis (i.e. if the parent cell produces daughter cells directly). In multicellular organisms only a small part of the organism is dedicated to reproduction. In these, there is local cell division. Occasionally, complete multicellular individuals result from vegetative propagation. A well-known example of this kind of reproduction is the creation of buds on the leaf-rims of *Bryophyllum* plants. But often the propagative units that result from mitosis are either single-celled or consist of very few cells. These are called spores (mitospores, to be exact). In principle, the mitotic propagative units from vegetative reproduction are all the same.

Generative propagation requires that individuals of opposing sexes are present in the same generation. This type of reproduction is unique to eukaryotes. Two cells (in actual fact, the nuclei) with different sexes fuse (syngamy). When the gametes combine and become a zygote the number of chromosomes is doubled. Two haploid ($1n$) cells produce a diploid ($2n$) product. During the following ontogeny the zygote becomes a complete diploid organism. This produces, at a later date, more generative reproductive cells. One could conclude that in successive generations the number of chromosomes doubles, but this is not the case. In fact, just before gametes are produced the chromosome number is halved. This process, in which the number of chromosomes is reduced to the original number of chromosomes, is known as *meiosis*. It is similar in some aspects to mitosis, but it is much more complex.

During meiosis two nuclear divisions occur, one after the other. In contrast to mitotic divisions, during the first meiotic division entire chromosomes (as opposed to chromatids) end up in each of the daughter cells. The matching chromosomes from the female and male cells separate, then recombine at random. A further rearrangement occurs when the two chromatids from the maternal and the two chromatids from the paternal chromosomes exchange parts of their DNA in a process called *crossing over*. Crossing over is an enzyme-catalysed DNA recombination in which the DNA strands are broken locally and the broken parts of the male and female chromatids are exchanged. By retaining its species-specific structure, this extensive new combination of the parental genomes is another aim of meiosis which is at least as important as the reduction of the chromosome number. At the end of the first meiotic division the chromatin recondenses only slightly and a new membrane forms only temporarily. Until the second meiotic division the cells will stay in this intermediate state (*interkinesis*). In contrast to the inter-phase during mitosis, there is no DNA replication at this stage. After a further condensation of the chromosomes and dissolution of the nuclear membrane, the second mei-otic division occurs which is, in principle, identical to mito-sis. The result of the two meiotic divisions is that one diploid cell produces four haploid nuclei, which are called *meiospores* after cell walls have formed around them.

Just as the diploid zygotes of some plant species are able to further subdivide by mitosis and build a multicellular haploid organism, so can the haploid meiospores of other plants. In extreme cases meiosis occurs immediately after a zygote is produced. Such organisms, where only the zygotes are diploid, are called *haplonts*. Some green algae have haplontic life cycles. At the other end of the spectrum are diplontic organisms in which meiosis is initiated just before gametes are produced and which immediately form the zygote. This is not only the case with humans and other animals but is also known in some green algae and diatoms. In the plant kingdom the two extremes of haplon-tic and diplontic organisms are the exception rather than the rule (Figure **2.2a–d**).

The duration and relative contributions of the haploid and diploid phases vary considerably in different organ-isms. Diploid organisms always form haploid spores by meiosis. Except in the extreme case of a diplont, these spores do not function as gametes but divide by mitosis and develop into a haploid multicellular organism. This is called a gametophyte and it is analogous to multicellular diploid sporophytes that arise from zygotes. Such life cycles with two ploidy stages and a varying mode of propagation are known as alternating cycles. The two phases of ontoge-ny, which start with either a vegetative or generative germ cell, involve one or more mitotic cell divisions and which end in the production of germ cells, are called generations. Normally there are just two generations, which are propa-gated in different ways. The gametophyte starts from a germinating meiospore and produces gametes. The sporo-phyte starts as a zygote and produces meiospores. The generation that dominates the life cycle (on a time scale or

Figure 2.2 Various types of life cycle. The haploid phase (1*n*) is characterized by a single circle, the diploid phase (2*n*) by a double circle. Arrows mark the beginning and end of the phase. S, fusion of gametes (syngamy), M, (meiosis, reduction division). (**a**) Haplont—the diploid phase is limited to the zygote; (**d**) diplont—syngamy occurs immediately after meiosis; (**b**), (**c**) alteration of generations with the gametophyte (**b**) or dominant sporophyte (**c**).

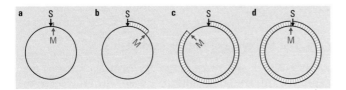

by cell number) may be the gametophyte (for example in mosses). However, in the higher vascular plants the sporophyte dominates. Both generations may either be very simple or develop into extremely complex organisms. The following section gives characteristic examples of the various groups and should reveal the basic principles.

2.2.2 Simple life cycles

Prokaryotes and single-celled eukaryotes develop in the simplest fashion—they propagate vegetatively by binary fission. For example, the single-celled, ovoid green alga *Chlorella* produces *autospores* which differ from their mother cell in size only. These divide and complete the cycle. Other single-celled green algae multiply generatively. For instance, the alga *Chlamydomonas* produces gametes spontaneously when exposed to an unfavourable environment. The flagellated gametes consist of + and − strains. Two of these fuse to form the zygote. After a growing phase, meiosis begins and the four zoospores become the vegetative haploid form of the alga. The only diploid stage during their development is the zygote. *Chlamydomonas* is therefore a haplont. External conditions and circumstances determine whether it reproduces sexually or asexually. Other algae change regularly between vegetative and generative reproduction. This is a genuine alternation of generations. This is exemplified in the siphonous green alga *Acetabularia* which has an important role in developmental and cytological research (section 2.4.2).

All kinds of life cycles can be seen in various algal groups. After the plants conquered the land their development became based upon the same life cycle: mosses, ferns, gymnosperms, and angiosperms all have readily distinguishable gametophytic and sporophytic phases. The sporophytes are usually more prominent. In mosses, the gametophyte is the actual plant, in ferns it is reduced to a small, leaf-like body, the *prothallus*. Seed-producing plants have reduced the gametophyte to such an extent that it is no longer recognizable as a separate entity. Different names have been given to the same cells and tissues of different vascular plants, even though they have the same function in the life cycle. The comparative terminology is beyond the scope of this book and details should be taken from a botany textbook.

2.2.3 The life cycle of higher land plants

The egg and a sperm cell fuse to form the zygote (the beginning of the sporophytic phase) from which the embryo develops. After a characteristic sequence of embryogeny with complex modifications (*see* section 6.1) the embryo lies in the seed (that is, the embryo, with its surrounding endosperm and seed coat). It can survive adverse conditions and disperse easily. When the seed germinates, the seedling emerges as a simply organized but fully functional young plant. Continued growth will result in the fully functional adult plant. At this stage the developmental cycle switches from the vegetative to the generative phase. The visible sign is the development of flowers (*see* section 6.3). Meiosis occurs within the generative male and female flower organs. The micro- or macrospores (here called pollen grains or embryo sac cells) germinate to form either micro- or megagametophytes. The latter are extremely reduced structures (just a few cells) inside the flower organs. Here the gametes are formed—either sperm or egg cells. After fertilization a new sporophytic phase of the generation cycle is started (Figure **2.3**; *see also* section 6.3).

In higher plants the development of the gametophytes and embryogenesis take place within the flower and are not visible from the outside. Here sporophyte development is dominant, and its highlights are seed germination and flower formation. In annuals the onset of generative development coincides with the beginning of a general degradation.

The different phases of development in the two generations can be defined and provide a useful set of characters to apply to plant development. Thus one can distinguish between the main vegetative and reproductive states. The first of these can be further divided into embryogenic, juvenile, and adult phases. Other phases of development describe events such as seed germination and flower development. In many instances the state is less obvious (i.e. between juvenile and adult leaves of ivy). Another justified distinction may be the differences between active and resting phases. During a plant's life active phases regularly alternate with apparently static phases during which there is no visible external development (i.e. seed maturation, seed germination; *see* section 5.6).

2.3 Key processes of development: growth, differentiation, and morphogenesis

It is useful to separate the developmental events into several individual components: growth, differentiation, and

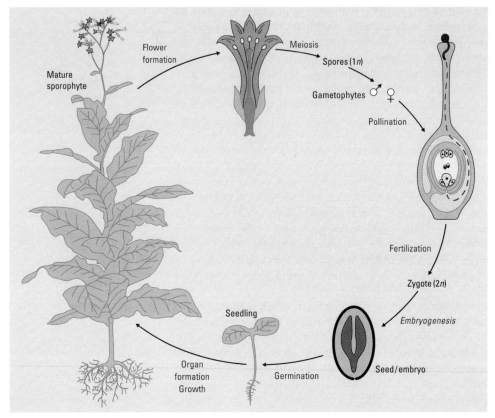

Figure 2.3 Life cycle of an angiosperm plant (example: *Nicotiana tabacum*). The diagram illustrates the characteristics of the life cycle: the dominating sporophytic phase lasts from the formation of the zygote (bottom right) to the mature flowering plant (top left). The principal phases in sporophyte development are embryogenesis and seed maturation, germination and development of the seedling, growth to the mature plant, and finally flower formation. After this the fruit and seed formation phases follow, which are already characterized by senescence, and sooner or later end with the death of the plant (not shown). The gametophytic phase is restricted to events in the flower (upper and top right of the drawing). At the beginning of the gametophytic phase meiosis produces the spores (formation of the pollen grains from the pollen mother cells in the anthers and formation of the embryo sac cell and the embryo sac mother cell in the ovule, *see* section 6.3). These micro- and macrospores form the extremely reduced, few-celled male and female gametophytes with their generative cells (the germinated pollen grain with sperm cells or embryo sac with egg cell; *see* section 6.3). The gametophytic phase ends when the gametes fuse to form the zygote and the cycle starts again. (According to Goldberg 1988.)

morphogenesis (Figure **2.4**). The term *growth* is used to define the gain of volume and/or mass of an organism due to cell enlargement and cell multiplication. *Differentiation* describes alterations of the form and function of cells, resulting in the formation of new structures or changes in tissues and organs. The various special functions of cells determine the appearance of multicellular organisms. The changing shape of the organism which is linked to changes in function, is called *morphogenesis*. Differentiation and morphogenesis may describe the same phenomena but according to the above definition differentiation relates to cells and organelles and morphogenesis describes even more complex events at the level of organs or the entire organism (Figure **2.4**).

2.3.1 Growth: the increase of size and weight

A size increase of the whole organism is easily recognized and external dimensions such as diameter, height, volume, and weight can be measured. After homogenization and extraction, the biochemical properties of the plant material may be established. Some of these (such as pigment content, enzyme activity, and protein, RNA, and DNA content) can be used to characterize the organism. However, these measurements are the sum of the activities in an entire organism. As plants consist of many different organs and tissues, which are composed of many cells, one has to question whether the externally visible and biochemically measured growth reflects that of the whole organism or of just single parts. Single cells show two major components of growth: division and enlargement.

Division of cells

Every cell has its specific developmental cycle, the *cell cycle*. This cell cycle, which is normally completed in a few days, shows distinct phases: a longer phase with intensive synthesis and decondensed chromatin (the *interphase*), and a shorter phase with condensed chromosomes during which nuclear and cell division is apparent (M phase; section 2.1). Interphase may be divided into three observable subphas-

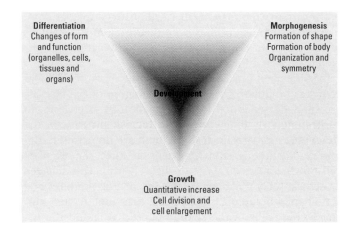

Differentiation
Changes of form
and function
(organelles, cells,
tissues and
organs)

Morphogenesis
Formation of shape
Formation of body
Organization and
symmetry

Development

Growth
Quantitative increase
Cell division and
cell enlargement

Figure 2.4 Growth, differentiation, and morphogenesis—the three basic features of development. The three terms are defined.

es: during the S phase DNA replicates; in the preceding G1 and following G2 phases, DNA is not replicated (G = gap) but RNA and proteins are synthesized. The M phase can be divided into *prophase, metaphase, anaphase,* and *telophase.* These divisions are based on the condensation and characteristic arrangement of the chromosomes and other cytological features not discussed in detail here. The duration of the cell cycle and the orientation of the cell divisions have an obvious effect on the developing organs and the final shape of the entire organism. This is reflected in the term *formative growth,* which contrasts with *proliferative* (mass-gaining) growth.

Cell enlargement

In plant cells there are two different mechanisms of volume and mass growth: increases in biomass and enlargement by stretching. Plant cells gain most of their additional biomass after mitosis when the cell volumes of the two daughter cells increase again to the initial volume of their dividing mother cell. Stretching does not necessarily involve a general increase of organic substances: frequently the amount of cytoplasm remains virtually constant or decreases slightly. Volume increases of the whole cell are due to intake of water (vacuoles are formed); this will be accompanied by stretching and surface growth of the cell wall. Stretching is not omnidirectional; expansion is usually limited to one direction. This is why it is referred to as growth by stretching.

Plants cells that show this selected stretching are, for example, the elongated (*prosenchymatic*) cells that are used to transport substances and which contribute partially to the rigidity of the plant structure. The top values for speed and duration of elongation are achieved by the cells of the bamboo shoots, which stretch with an average speed of 0.4 mm/min (more than 50 cm/day) over several days. Even faster elongation has been measured in the stamens of rye (2.5 mm/min); however, here elongation is completed after 10 min.

In some plant organs (for example the roots) the cellular activities of growth by division or addition of the biomass are spatially separated from elongation by stretching. In roots these two zones can be distinguished easily. In the shoot apex the two modes of growth are not easily recognizable. Frequently, stretching does not affect the complete cell but is limited to parts of the cell wall. This selected growth in several regions on the cell surface produces odd cell shapes. Examples of this are the single-celled alga *Micrasterias,* cells of the spongy parenchyma and the trichomes (hair-like epidermal cells) of higher plants (*see* section 6.2.7). The underlying principle of differentiation from a meristematic cell (which is able to divide) is the development of specialized cells which are no longer able to divide.

2.3.2 Differentiation: specialization of form and function

Growth deals with quantitative changes during development, differentiation describes the qualitative changes. Ontogenesis results in morphological differences which coincide with a specialized function.

Definition Differentiation involves changes of the morphology and function of cells, tissues, and plant organs.

One can recognize the changes even in organelles. A very well known example of this is the differentiation of plastids (i.e. the transition from proplastid to chloroplast; *see* Box **4.2** and section 5.6.5). This differentiation is visible externally because specific membrane structures and the photosynthetic pigments (greening) are formed. However, one has to keep in mind that the differentiation of organelles is only part of the development that is controlled by the complete plant cell. The term differentiation is therefore often equated with cell differentiation. While differentiation is often limited to a single cell, which develops distinctly from all its neighbours (as an *idioblast*), normally groups of cells or specific regions are affected; this is known as tissue or organ differentiation.

Typically, differentiation leads from the meristematic cell to cells with special functions which cannot divide. However, differentiation may also proceed in the opposite direction (called re-embryonalization—closure of wounds with newly formed callus cells). Callus tissues, isolated cell cultures, and protoplasts (which lack cell walls and are made in the laboratory) may be induced to renew differentiation. From these, specialized cells with completely different structure and function may be formed. Single cells can be stimulated to regenerate to complete plants (protoplasts from mesophyll cells are often used), this is referred to as the *totipotency* of cells and reflects the fact that single, completely differentiated plant cells can retain all necessary information for the development into any specialized cell type. Normally, fully differentiated cells first form meristematic cells which consequently re-differentiate. However, there are known exceptions; for example, in tissue cultures the isolated mesophyll cells of *Zinnia elegans* differentiate directly to form specialized cell shapes (the *tracheary* elements).

2.3.3 Morphogenesis: developing the form of the complete organism

Cell and tissue differentiation results in a typical and recognizable external appearance of the organism. However, the basic form of the body is not constant but changes during the developmental process. One only has to compare the various repeated or even identical parts of the organism to discover divergent forms or alterations in symmetry. During plant ontogenesis we can find all kinds of symmetries: the spherical early embryo has a radial symmetry; with further embryogenic development this symmetry becomes bilateral—the two halves of the plant are mirror images. This is particularly evident during the formation of two cotyledons in angiosperm embryos. However, there are obvious differences between the apical and the basal halves of the embryo which form the shoot or the root precursors (*see* sections 6.1 and 6.2). During the development of the flowers (but not of the embryos) we often see dorsoventral symmetry. Here there are differences on the upper and the lower sides which are defined by one of the two possible mirror images of the flower on the longitudinal axis. Metameric symmetry is a third type, used to group the events where elements along an axis are formed at the same distance and have similar orientations with respect to one another. Well-known examples are pinnate leaves and the scales that cover pine cones. Special forms develop when the axis is a spiral or a helix. Any combination of basic forms having two types of symmetry produces new overlapping symmetries that are very complex. The whorled arrangement of leaves on the stem illustrates these complexities. The individual whorls are radially symmetrical with equally spaced leaves. Additional complications arise from the spatial arrangement of the neighbouring whorls on the shoot axis which are based upon metameric symmetry (*see* section 2.5.2).

Morphogenesis summarizes the visible expression of a number of highly complex processes (section 2.2.3). It includes newly formed structures (for example that of the embryo from the zygote) and changes in form and function (i.e. seedling development, the transition from seedling to adult plant, the formation of flowers, seeds, and fruits, and senescence in annual plants).

Embryogeny in plants follows a predictable pattern of cell divisions. This kind of homeostasis leads to species-specific developments in which every step towards the final appearance is controlled by endogenous regulators.

The crossing of plants and analysis of mutations have revealed that endogenous regulation of development resides in specific genes. However, it is important to note that the endogenous regulation of embryogeny in angiosperms is somewhat variable in the different species.

In contrast to embryogeny, which is largely regulated by endogenous factors, the other main phases of plant development are strongly influenced by additional environmental factors. The development of angiosperm seedlings in the light (photomorphogenesis) or in the dark (scotomorphogenesis) (*see* section 5.2.1) is one example. Another is light- and temperature-induced flowering in plants (*see* section 6.3). One of the main goals of plant developmental biology research is to understand these interactions between internal and external factors.

Pathological morphogenesis

Laboratory experiments can induce, and spontaneous natural events may result in, morphogenic processes that are not part of the normal development of the individual. These are, for example, the formation of tumours and galls. Apart from their practical importance, much can be learned from these processes because they show the extremes of plant development.

Tumours

Malignant tumours (cancers) are caused by uncontrolled cell growth in humans and animals. Plant tumours are also characterized by the loss of normal tissue differentiation and a simultaneous uncontrolled division of the affected cells. However, it is important to note that tumours in plants are very different from those in animals as they require an induction by special plant viruses and bacteria, which will be discussed in Chapter 7.

The plant tumours which are caused by infections are different from the so-called *genetic* tumours. The latter are formed by uncontrolled tissue growth in interspecific hybrids of *Brassica, Datura,* and *Nicotiana* species and do not occur in the parental plants. Similar tumours are also known in the animal kingdom (in *Drosophila*; melanomas in life-bearing tooth carps and other poeciliid fishes). It is thought that the genetic tumours of hybrids may be due to a partial incompatibility between the two parental genomes. As a result of this incompatibility the flow of cellular genetic information is disrupted and may, in extreme cases, be expressed as the tumour phenotype. Regeneration experiments with tumorous tobacco hybrids have shown that tumour cells have a complete set of genes and that the next generation has tumour-free cells. This feature is the fundamental difference between plant and animal tumours.

Galls

Penetration of leaf tissues by insects (certain wasps, mosquitos, or mites) and the subsequent deposition of eggs produces a multicellular gall which surrounds the developing insect larvae. These galls are spherical and have an internal organ-like pattern. The morphogenesis and phenotype are species specific. Galls are not induced by a single environmental stimulus (i.e. tissue injury or deposition of eggs) but require the constant supply of morphogenetic substances, which are produced by the developing larvae. This feature distinguishes real galls from the crown galls caused by *Agrobacterium* infections.

2.4 Reasons for cell differentiation

2.4.1 Chemical, electrical, or mechanical signals?

As we have seen, cell differentiation can happen at many levels, and ranges from macroscopic changes in shape via microscopic and submicroscopic changes of internal structures, to biochemical changes. Biochemical differentiation involves the vast spectrum of metabolic products, the metabolites and macromolecules. The current state of the art suggests that this is the level that will provide many answers about cell differentiation. However, one has to consider other possible contributors, including chemical substances with low molecular weight (i.e. phytohormones; *see* section 5.5) or electrical and mechanical stimuli. For example, experiments with protoplasts have shown that cell surfaces have a particular electrical charge which is influenced by the surrounding media. Plant cells in tissues are subjected to forces generated by neighbouring cells, which depend upon the degree of vacuolization and the type of cell wall. These chemical and physical influences have been largely ignored as contributors to cell differentiation. Their possible involvement and role during differentiation is still to be investigated.

2.4.2 Genes and gene expression

The important role of genes and their effects on cell differentiation have been studied in many mutants that lack defined developmental traits (*see* Chapters 3 and 6). The classical studies that provided proof of the role of nuclear genes were conducted with siphonous algae of the genus *Acetabularia*.

Acetabularia looks like a single-celled organism and is ideal for experimental manipulations. The large primary nucleus can be removed easily and injected into stalk segments. The nucleus will not only programme the stalk for its own growth and development but also induce its species-specific cap (the whorl of gametangial sacs at the top of the organism), providing conclusive evidence that the nuclear genes are essential. If the rhizoid of *Acetabularia crenulata* (it has a small cap), which contains the nucleus, is transplanted into the stalk segment of *Acetabularia mediterranea* (which has a large cap), the smaller *A. crenulata* cap forms. The alternative combination induces the larger *A. mediterranea* cap (Figure **2.5**). Similar results are obtained by implanting isolated nuclei directly into stalk segments of other species. Initially, the specific cap formation was thought to be due to morphogenetic substances that are released from the basal primary nuclei into the cytoplasm to influence the formation of the cap in the apical stalk area. Investigations with inhibitors that prevent the synthesis of RNA and proteins suggest that morphogenesis of the cap is dependent upon species-specific RNA molecules which are translated into the required proteins.

2.4.3 Differential gene expression: from nucleotide sequence information to time- and space-dependent cell activities

The observations of *Acetabularia* cited above demonstrated the importance of the nucleus for cell differentiation. Using current resolution techniques, genes and their products

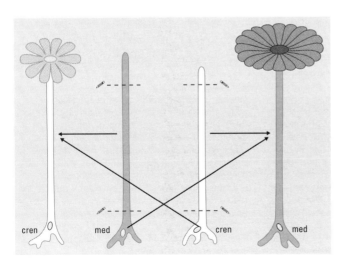

Figure 2.5 Transplantation experiments in *Acetabularia mediterranea* and *Acetabularia crenulata*. The influence of the nucleus during cap formation is demonstrated when young stalk segments are grafted on to rhizoids that contain a nucleus of the other species: the nucleus in rhizoid *A. mediterranea* (*med*) causes *med* type cap formation on stalk segments of *A. crenulata* (*cren*). A *cren* rhizoid leads to the formation of a *cren* cap on a *med* segment. (Adapted from Hämmerling 1963.)

can be visualized, permitting comparative studies of the total complement from different cell populations. Synchronized cultures of algae are often used as experimental systems (section 2.2.2) because it is easy to obtain large amounts of uniform material for investigations. One can also isolate various specialized cell types from tissues of higher plants; for example, the two cell types that can be isolated from the so-called C4 plants by treating the leaves with a mixture of enzymes that degrade cell walls (these mesophyll and vascular bundle sheath cells differ in their RNA and protein content; *see* Box **4.4**). The already-cited formation of tracheary elements in isolated mesophyll cells of *Zinnia* has also been linked with genes that induce different RNA and protein expression patterns. With the available techniques, it is not even necessary to isolate the cells if one wants to examine cell-specific differences between the activities of certain genes. *In situ* techniques display gene products in fixed cells and tissue sections. These widely used techniques permit insights into the mol-

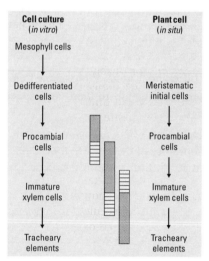

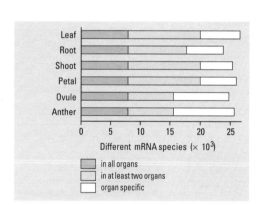

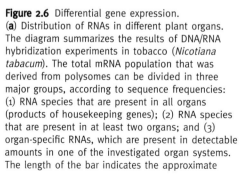

Figure 2.6 Differential gene expression.
(a) Distribution of RNAs in different plant organs. The diagram summarizes the results of DNA/RNA hybridization experiments in tobacco (*Nicotiana tabacum*). The total mRNA population that was derived from polysomes can be divided in three major groups, according to sequence frequencies: (1) RNA species that are present in all organs (products of housekeeping genes); (2) RNA species that are present in at least two organs; and (3) organ-specific RNAs, which are present in detectable amounts in one of the investigated organ systems. The length of the bar indicates the approximate

number of various mRNA types in the total mRNA population (adapted from Goldberg 1988). (b) Differential gene expression during the formation of tracheary elements in *Zinnia*. Young seedlings form precursor cells. In mesophyll cell cultures under appropriate conditions spontaneous differentiation occurs, with intermediates that resemble those formed in intact plants. Both in the cell culture system (*in vitro*) and in tissue sections (*in situ*) RNA species can be detected that accumulate during certain phases of tracheary element differentiation (see bar). (Adapted from Demura and Fukuda 1994.)

ecular state of differentiation of the tissues of an organism (Figure **2.6a,b**).

During the transition of a cell from differentiation state A to differentiation state B the activity of a single gene may not be influenced. It is also possible that its activity will be either higher or lower with regard to the total measurable activity of the genes. The quiescent genes, which are by far the largest group, include the so-called *housekeeping genes*, which are responsible for the enzymes involved in basic cellular metabolism. This group is bound to contain genes that are not influenced during a transition from A to B but may be involved in other differentiation activities, for example from A to C or from B to C. Expression of one particular gene can increase during the transition from A to B and decrease in the transition from B to C. Also, the main level of regulation during formation of the gene products can differ from one situation to another; for example, gene expression can be regulated primarily at the level of transcription, during RNA processing, or while the mRNA is translated (*see* Chapter 4). If there is positive regulation at one level of gene expression and negative regulation at another, the amount of gene products will even out. This example shows that the terms *gene activity* and *differential gene expression* are purely descriptive and do not reveal anything about mechanisms. These can vary during the developmental cycle and depend on the experimental system (species, tissue, or cell type). Every set-up will require its own additional analysis.

It has to be pointed out that the proof for differential gene expression is only a first tentative step towards a causal analysis of cell differentiation. Further related questions arise here; for example, how is information in the sequences of biological macromolecules incorporated into complex cell structures or used for co-ordinating cell activities? In addition, there is the question of causality: are the changes in the gene expression pattern during the transition from A to B the effect or the cause of cell differentiation? Indeed, the first observed genes with cell-specific expression are normally target genes for regulatory proteins. These regulatory proteins are also products of regulated gene expression. We often find these cascades of genes that are regulated by the products of the previous level.

2.4.4 Cell polarity

When we looked at cell division (section 2.3.1) we assumed that both daughter cells are genetically identical at the same state of differentiation. However, this is not always the case. If the maternal cytoplasm is not evenly distributed, the two nuclei in the daughter cells have different cytoplasmic environments. This may, for example, result in differing complements of plastids and mitochondria and may lead to differential gene expression patterns in the two daughter cells and their offspring. These dissimilar cell divisions are very important for the development of almost all groups of organisms. Examples in the plant kingdom include the first division of the zygote, and the formation of pollen grains, stomata, or trichomes. The uneven distribution of cytoplasm is circumscribed by the term *cell polarity*, i.e. the composition and the features of the cytoplasm change from pole to pole along a polarity axis. This polarity may be permanent; however, in most instances there is a phase during which cell polarity can be changed before it is aligned permanently. Such a change of cell polarity can be induced by external factors (i.e. light or electrical fields) but only as long as the polarity axis is not fixed endogenously.

Direct experimental investigations of cell polarity in higher plants is extremely difficult, therefore systems with single or only a few cells are used. Examples of these are spores of the genus *Equisetum* (horsetails), gametophytes of the fern *Dryopteris*, and protonema of the of the moss *Funaria*. Zygotes from the brown algal genus *Fucus* (Figure **2.7a–e**) have been particularly rewarding. These relatively large cells (100 μm in diameter) are initially apolar but form a polar axis 4–10 h after fertilization which can still be influenced experimentally. This axis becomes fixed 10–12 h after fertilization. It is visible as a localized protrusion and indicates the place of rhizoid initiation. After a further 12 h, the first cell division occurs, perpendicular to the polar axis of the zygote. This asymmetrical cell division forms a small rhizoid and a larger thallus cell. Further characteristic cell divisions produce the initially spherical thallus, which becomes lobed later, and the branched rhizoid. Physiological and biochemical investigations suggest that the cell wall contains at least some necessary information for cell polarity in the *Fucus* zygote. Certain cell wall polysaccharides (sulphonated fucanes) and a surface glycoprotein (vitronectin-like protein; section 2.5.3) have been found exclusively in the walls of the outgrowing tip of the rhizoid pole. How much these substances affect the induction of polarity and whether these findings can be extrapolated to higher plants remains to be established.

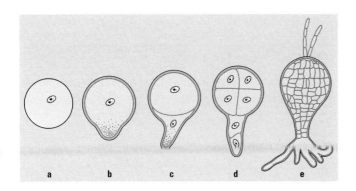

Figure 2.7 Cell polarity during development of the *Fucus* zygote. (**a**) In the fertilized egg cell there is no visible polarity. Within a couple of hours, the zygote forms an axis of polarity. Light and other environmental factors are important during this process. (**b**) Approximately 10–12 hours after fertilization the polar axis is fixed. On one side, polar growth starts with the formation of the rhizoid tip (first visible marker of cell polarity). (**c**) After another 12 hours the first cell division starts. The wall forms perpendicular to the direction of growth of the rhizoid tip. This results in two daughter cells that differ in size: the bigger thallus cell and the smaller rhizoid cell with the growing rhizoid. (**d**) Both cells divide again parallel to the plane of the first division. The two cells that emerged from the thallus cell divide perpendicular to the axis. (**e**) Further longitudinal and perpendicular divisions occur and result in a multicellular thallus with multiple rhizoids, some of which branch. (Adapted from Goodner, B. and Quatrano, R. S. *The Plant Cell*, **5** (1993), 1471.)

2.5 Regulation of development in multicellular organisms

Differentiation and morphogenesis have temporal and spatial components. The changes that lead to the differentiation of cells and tissues and eventually to the expression of a pattern do not happen at random but they follow a predetermined plan within certain limiting boundaries.

Definition Pattern is the non-random distribution of structures in individual organelles or in the organism. Pattern formation therefore describes the spatial organization of differentiating parts to a higher level of organization. This system is organized and has a form.

In pattern formation, two separate phases are defined: an initial pattern specification and a final pattern realization. This implies that patterns are predetermined and inherently present before they are visible externally. This normally irreversible factor is called *determination*.

Definition Determination is the commitment of cells along a particular and virtually irreversible path to differentiation.

The list of developmental biological terms is not complete without the term *competence*, which describes the ability of the cell, the organelle, or organism to complete a specific section of development (Figure **2.8**).

At first glance this terminology of developmental biology seems to be fairly strange. As in many other areas of science, the expressions were chosen before many of the complex features and the underlying molecular details were understood. Today this terminology provides a known sound foundation of knowledge which can be added to by information provided by the use of advancing techniques in molecular biology.

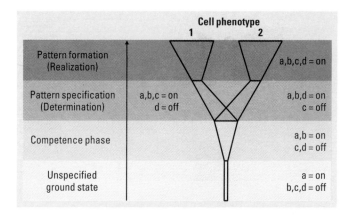

Figure 2.8 A model for cell determination. The left-hand side shows the consecutive phases of increasing manifestation of a specific cell phenotype. On the right-hand side some of the complicated developmental events in cells have been reduced to activation ('on') or inactivation ('off') of a small number of genes that act as hypothetical switches (a, b, c, d). The activity of these genes produces regulatory proteins (factors) which are active during certain phases and may be necessary for commencing the next phase of cellular development. This scheme assumes that in the early state (unspecified ground state) only one of the four depicted genes ('a') is active. Only after the activation of a second switch gene ('b') a transition into the next developmental phase occurs. Passing through this phase is important for the later expression of a specific cell phenotype (the competence phase). This phase broadens the developmental potential. Thereafter development may be narrowed into different directions and depends on the activation of the switch genes ('c' or 'd'). This is the phase of pattern specification. After programming into a certain developmental direction has taken place, that path will be followed (phase of pattern formation). At this stage it would not make any difference which factors are missing ('d' or 'c'). The previous timing of their production is important as this enables the balanced interplay with other developmental genes and their products.

2.5.1 Determination: the commitment to differentiation fates

We have seen in section 2.3.2 that a cell isolated from a tissue has a broad developmental potential. This is evident during its regeneration to an intact plant, when a large number of diverse daughter cells form from this single cell. However, this does not mean that every cell in a tissue can develop just as it pleases. Just the opposite is true as the organism controls the development and the direction of differentiation of the individual cell. This kind of mechanism is essential for the development of a species-specific plant body. It is achieved by imposing increasing directional growth limitations on a plant cell. This may be demonstrated by the rolling ball model (Figure **2.9a,b**). In the epigenetic landscape the ball moves slowly down a gentle slope and can take several possible paths. The ball may cross the gentle barriers or roll back if it has not got sufficient energy and/or pursue a different path. With

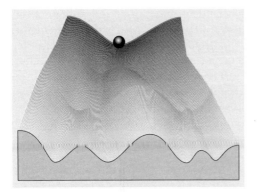 

increasing barrier heights and steeper slopes the descent will be governed by reduced choices until there is only a single path left which leads downwards.

There are quite a few reasons why this model is much too simple for plant cells:

1. In a tissue there are no real barriers. Plant cells have the ability to change their differentiation patterns even when development seems firmly entrenched.
2. Sections 2.5.4 and 2.6.3 will demonstrate that differentiation of the cell is very much more dependent upon its location within the tissues of an organism than on information in an ancestral cell.
3. The mechanical rolling ball model has fixed parameters but the fate of plant cells during differentiation is determined by a spectrum of additional internal and external factors (such as light, temperature, etc.; *see* Chapters 5 and 6). Plants are rather more flexible during their development than the model that has been chosen to illustrate the principles of determination.

Nevertheless, the rolling ball model is a very useful tool for elucidating determination. We can assume that a plant cell or an organelle is following a particular direction if reversal of its development is difficult or impossible.

This definition shows that most plant cells, apart from zygotes and early embryos, have a determined development. This often holds true for callus cells, which may be determined to form roots, shoots, or undergo somatic embryogenesis. If one establishes callus cultures from juvenile and mature *Hedera helix* (ivy), the two types of callus differ morphologically and physiologically. The cells of the juvenile callus grow faster and are larger than those of the mature callus. Moreover, the juvenile callus regenerates to form juvenile plants and the mature callus to mature plants.

Figure 2.9 Model of a rolling ball in an epigenetic landscape. (**a**) The cell is shown as a ball in a three-dimensional landscape. Progressive narrowing constrains the ball from several to a single possible path. (**b**) The selected path is determined by the profile of the epigenetic landscape. This profile is formed and influenced by underlying poles and ropes (genes and gene products) that are woven into a regulatory network. (According to Waddington.)

We can assume that determination of cells follows a certain differentiation route during which the participating cells change their molecular developmental repertoire step by step and are able to re-route the direction. At the outset their differentiation is probably reversible. Before an irreversible state is attained the cells are likely to pass an indifferent phase in which the plant is capable of changing to new developmental directions. This capability is called *developmental competence*. To date competence and determination have not been defined at the molecular level. It is evident that these are not clearly separable developmental processes with switch-on and switch-off mechanisms; instead they have to involve networks of information and interactions which form an ever-increasing number of meshes which finally determine differentiation. It is interesting to question whether the characteristics of the cell are permanent and whether these are passed on in future cell lineages (section 2.6.3).

2.5.2 Pattern formation

The cellular structure of plants is due to the fact that many related cells within a developing tissue or plant organ follow simultaneous developmental steps. However, comparable local cellular differentiation is frequently separated in space and/or time. Also, neighbouring cells in a tissue can adopt different developmental strategies. This leads to morphological and physiological differences within developing organisms which form externally visible patterns. Examples of such patterns are the ordered arrangement of leaves (*phyllotaxis*) and the branching of roots and shoots.

Besides these patterns in plant organs there are also examples of cellular patterns. The distributions of stomata and trichomes in the epidermis belong to this type of pattern. These all have one feature in common: their individual components are highly structured. It has been suggested for a long time that these patterns are dependent upon links between the individual components. A very popular hypothesis suggests the involvement of inhibitory factors. The presence of controlling influences may be demonstrated in leaf primordia. When one of these primordia is damaged or removed the position of the remaining leaves will change.

The causes of pattern formation are still largely unknown. One important clue may be the observation that patterns always start with asymmetrical cell divisions. This can be studied during the development of stomata, which are important for the exchange of gases and osmoregulation by the plant but vary in their morphology in different plant groups (Figure **2.10a,b**).

Stomatal development starts with an asymmetric division of a single epidermal cell. The small, cytoplasm-rich mother-cell is the guard cell precursor. The other cell develops into a normal, highly vacuolated epidermal cell. In many instances the neighbouring epidermal cells also divide unequally. The smaller cells surround the mother cell and are called subsidiary cells. The stomatal complex is completed after a symmetrical division of the mother cell, producing the two *guard cells* which form the stomatal aperture.

2.5.3 Cell–cell contact

Pattern formation is closely linked to the question of communication between cells, their neighbours, and their surroundings. In animal cells, which lack cell walls, surface adhesion plays a key role. A couple of molecular mechanisms are known to ensure the correct arrangement of cells during development, including the recognition of the plasma membranes of neighbouring cells. The cell surface glycoproteins (*cadherins*) and components that interact with the extracellular matrix (glycoproteins such as

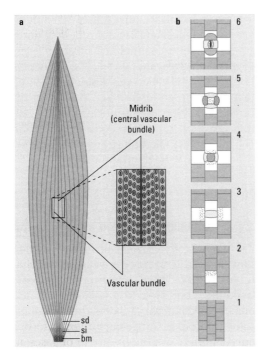

Figure 2.10 Development of stomata in *Tradescantia*. (**a**) Diagram of the leaf with midrib and parallel venation (monocotyledonous plant). The stomata are in the mature distal part, only on the lower side, of the leaf. They form a very regular pattern parallel to the veins (see enlarged section). In the basal part of the leaf a meristematic region (bm, basal meristem; section 2.6.2) is followed by a zone of stomata initials (si, stomata mother cell) and the area of stomatal differentiation (sd). Stomatal development is illustrated in (**b**): 1, undifferentiated (meristematic) cells; 2, formation of stomatal initials by asymmetrical cell division; 3, formation of the first pair of subsidiary cells by unequal division of neighbouring cells; 4, formation of the second pair of subsidiary cells; 5, formation of guard cells by unequal division of the stomata initials; 6, the mature stomatal complex is formed after cell elongation. (Modified from Croxdale *et al.* 1992.)

fibronectin or vitronectin), with receptors in the plasma membrane (integrin), are known adhesion mechanisms.

In plant cells the cell membranes of neighbouring cells are rarely in direct contact. Each cell is surrounded by a cell wall that consists of complex polysaccharides and proteins. Here adhesion is achieved mainly by neighbouring cell walls. For this reason isolated individual plant cells cannot aggregate. Aggregates that are normally observed in tissue cultures are due to cell divisions because the daughter cells adhere with the newly formed cell walls.

Specific interactions of plant cell surfaces do not only occur in one plant but also between cells of different plants (fertilization; *see* section 6.3.4) and very often between plant cells and other organisms (bacteria and fungi, for example). The best studies concern the events that lead to T-DNA transfer by *Agrobacterum tumefaciens* (*see* section 7.2) into the plant cell, the rhizobia/legume interactions (*see* section 7.3) and the pollen–stigma interaction within the flower (*see* section 6.3.4). In these and others there are highly specific interactions on the surface of the plant cell.

However, surface recognition is not the only method by which cells establish or maintain contact: the cell wall of a typical plant cell has pores that serve to connect the cytoplasm of neighbouring cells. These *plasmodesmata* facilitate a continuous intracellular space, called the symplast, in tissues. Plasmodesmata are not tube-like pores but provide a communication channel and a barrier. Plasma membrane covers their walls and, on the inside, there is a membrane-bound channel called the desmotubulus, which is in contact with the endoplasmatic reticulum of neighbouring cells. Therefore, communication between two cells via their plasmodesmata can occur in three different ways:

(1) via the cytoplasmic strand between the plasma membrane and the desmotubulus;

(2) along the plasma membrane; and

(3) within the desmotubulus (Figure **2.11**).

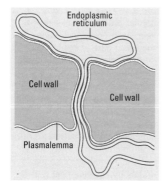

Figure 2.11 Plant cell connections via plasmodesmata. The endoplasmatic reticulum (ER) of neighbouring cells is interconnected by the central ER tube (desmotubulus) which passes through the plasmodesmatal channel. (From Nultsch W. *Allgemeine Botanik*, Thieme, Stuttgart (1991).)

2.5.4 Positional information and its interpretation

How does a cell in a tissue maintain its specific identity and how do neighbouring cells influence each other during differentiation? These questions provoked quite a few hypotheses and models from biologists interested in development. Most of these models propose morphogenetically active substances (activators or inhibitors) which are produced from a single cell or from groups of cells and are released into the surrounding environment. These substances can be particularly effective for pattern formation if they spread with a concentration gradient through a tissue. This gradient model proposes that a cell recognizes its location along a morphogen gradient and follows a specific differentiation path according to its position along the gradient. A further extension of this is the polar coordination model (Figure **2.12a,b**). Here it is assumed that the morphogen spreads radially, from a central point source with the highest concentration to lower concentrations at the periphery; every cell can be assigned its own radial and polar coordinates. Both models have in common that a disturbance of the morphogen gradient will lead to an adjustment of differentiation in involved cells; for example, the polar coordination model predicts that removal of tissue sections will induce cell growth and division until the original system of coordinates is restored.

A morphogenetic gradient can emerge spontaneously or may be present from the very beginning (i.e. the polarity of the egg cell). It is, however, more likely that a new gradient emerges from pre-existing patterns. This tightens the spatial networks of local information, drawing them closer and more intricately to each other. However, this tells us little about how these gradients emerge from the more or less homogeneous initial state. The so-called *diffusion reaction* model provides an explanation for this. This model is based upon two opposing processes: the autocatalytic production of an activator and a counteracting synthesis of an inhibitor, which in turn hinders the autocatalysis. In addition it is thought that this inhibitor diffuses faster than the activator. Mathematical calculations and computer simulations show that these conditions result in an inhomogeneous stable distribution in an area. This has local high and low values of the activator and the inhibitor (maxima and minima in an oscillator system). This model explains the formation of simple patterns but also the formation of repetitive periodic structures during growth. Typical examples which can be predicted accurately are regularly ordered structures along an axis (phyllotaxis on a shoot) or the formation of net-like structures (the pattern of the vascular bundles on the lamina).

What are these morphogenetic fields and gradients? If one accepts that they are caused by irregularly distributed molecular substances, then there are two principal methods for the adjustment of the gradients, either

(1) the morphogen can diffuse freely and serve as a direct signal for neighbouring cells or for more distant cells

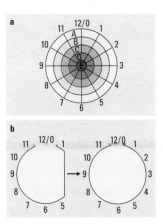

Figure 2.12 Polar coordinate model to illustrate positional information of cells. (**a**) A three-dimensional tissue section (i.e. the shoot apex) is shown diagrammatically as a projection from above. Each cell receives information about its position in this coordinate system, indicated by the letters A–E (highest point) and the numbers 1–12/0. (**b**) When the spatial structure is disturbed, for instance by injury, cell division and growth occur because the altered, and now 'wrong', coordinates in this area have to be adjusted until the starting point is reached again. (Adapted from French *et al.* 1976)

(for this, the signal would be a substance with low molecular weight); or

(2) the morphogen may be a molecule which, perhaps because it is too large to diffuse freely, can act by participating in an intracellular signalling chain or by signal transmission at the cell surface (section 2.5.3); this includes macromolecules among the morphogens.

During embryogenesis in animal cells inducers have been observed which function as gradient-forming signal carriers for cell pattern formation. These are, for example, different peptide growth factors (such as fibroblast growth factor (FGF) or transforming growth factor-β (TGF-β)) and non-peptide growth factors (i.e. retinoic acid and its derivatives). The diffusible carrier molecules interact with specific cell surface receptors. The cell's response to this signal is to synthesize mRNA and protein factors locally. This would be a local molecular interpretation of the information provided by the gradient. This cell-specific gene expression can lead to the formation of a new morphogenetic gradient and in turn facilitate a more exact molecular response according to position in the tissue. To date there are no known morphogenetic substances in plants but there is no reason to assume that the adjustment of morphogenetic fields or gradients should be principally different than in animals.

2.5.5 Long-distance effects

Plants are faced with the problem that they have to co-ordinate developmental processes between remote organs. Cell–cell transmission of signals would be slow and inefficient. For this reason long-distance communications between organs (*correlations*) rely upon the vascular system and frequently travel remarkable distances (for example in trees). This kind of communication may involve the long-distance transport of nutrients or plant growth substances with hormone-like characteristics (*see* section 5.5).

Long-distance transport of organic substances occurs mainly via the sieve tubes which are connected with the companion cells by numerous plasmodesmata. Sieve tubes and companion cells form functional units. All organic substances travel through them from their site of synthesis (source) to that of their consumption (sink). The direction of transport may be downwards (i.e. from the leaves to the roots) as well as upwards (i.e. from the roots or other storage organs to the flowers or the growing tips of the shoot).

Even though the transport of organic substances via the sieve tubes is important for transmitting long-distance effects, it should not be forgotten that electrical signals may function as transmitters from one organ to another. Young tomato plants respond with a measurable electrical impulse after the cotyledons are damaged. This impulse is thought to initiate a biochemical wound-healing response in the distant primary leaves.

Interactions between plant organs may enhance or inhibit developmental processes. An example of correlative enhancement is the promotion of cell division in the cambium of trees. Cell division starts here during the spring and is influenced by the developing buds. The signals are auxins and other phytohormones which are transported from the buds downwards. The growth of many fruits depends upon correlative enhancement by hormone transport from the ovule to the surrounding ovarian tissues.

There are many examples of correlative inhibition in which the development of one organ is negatively influenced by another; the best-known example is apical dominance. The tip of a plant or a branch of a tree inhibits the development of lateral shoots. If the tip is removed the growth of the lateral shoots is promoted. Applied auxin can imitate the inhibitory effect, and it is probable that this phytohormone at least helps to transmit this signal. However, other phytohormones may enhance the growth of lateral shoots in the presence of the apical shoot and act antagonistically. Here, too, the action of the hormones is likely to be an intricate balance between partially inhibiting and partially enhancing signals.

Section 2.4.4 introduced the term 'cell polarity' which was defined as an unequal distribution of material in the cytoplasm or on the surface of single cells in multicellular organisms. Many multicellular organisms may be cited as examples. They range from the generative and a vegetative pole in the colony-forming green alga *Volvox* to the higher plants, where at an early stage the poles for a root and shoot are formed (*see* section 6.1). Polarity in multicellular tissues often leads to a one-way transport of phytohormones. Here the mediators of polarity correspond to activities of those molecules involved in long-distance correlation. A classical experiment for the demonstration of this effect is the formation of lateral roots and shoots on branches removed from willow trees (Figure **2.13a–c**). These branches have buds along their entire length but the new shoots will only develop from the anterior buds. The

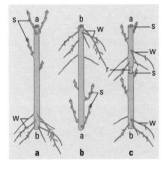

Fig. 2.13 Polarity in sprouting willow shoots. (**a**) Original; (**b**) inverted orientation; (**c**) with a ring of bark removed. a, (Original) apical; b, basal pole; r, removed bark ring; s, lateral shoot; w, root. (From Nultsch W. *Allgemeine Botanik*, Thieme, Stuttgart, 1991.)

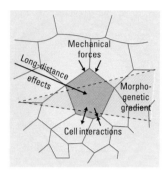

Figure 2.14 Influences on a cell in a tissue. The diagram summarizes the enhancing and inhibiting effects which have been discussed in the text: cell contacts by specific surface interaction and across plasmodesmata (section 2.5.3), morphogenetic gradients (section 2.5.4), and long-distance effects (section 2.5.5).

posterior region of the segment will form roots even if the segment is kept upside down. Phytohormones transmit the information over long distances through the phloem of the twig. If the bark is removed at any point on the branch, the regions around the wound will produce anatomically correct shoots or roots. This example shows that organ polarity includes correlative interactions based on long-distance transport.

We should keep in mind that long distance effects are not fundamentally different from the mechanisms that govern short-range development by direct cellular interactions (section 2.5.3) or morphogenetic gradients (section 2.5.4). It is important to remember that all these mechanisms are necessary for the regulation of cell differentiation (Figure **2.14**).

2.6 Development in plants and animals differs because plants possess meristems

2.6.1 Embryos

Even a brief glimpse at the animal kingdom shows that an almost complete but smaller version of the adult is present when the embryonic development is nearly complete. Vertebrate embryos have a complete body with the correct number of extremities and inner organs and their post-embryonic development consists mainly of further growth and completion of the existing organ systems.

Early separation into generative and somatic cells is typical of animal development. Generative cells retain their developmental competence and differentiate in the gonads to form the gametes. The strict separation of generative and somatic cells ensures that the complete genetic material and developmental potential of embryonal cells is passed to the next generation. In somatic cells, differentiation is frequently irreversible; well-known examples are the loss of the nucleus during the formation of mammalian erythrocytes and the loss of chromatin during the embryonic development of the worm *Ascaris*, as well as genome rearrangement mechanisms in the antibody-producing cells of mammals. Even though many animal groups retain the ability to regenerate tissues or organs partially or completely after wounding, the formation of a completely new organism depends on the fusion of two gametes to form a zygote. This suggests that only the gametes retain an intact functional genome, as well as the full cytoplasmic information for a complete developmental programme.

Plants do not have this separate germ line, i.e. generative cells that are set aside early during embryogenesis. Gametes are formed at a comparatively late stage in the life cycle and the cells that differentiate to gametes originate from the vegetative plant. This transition from the vegetative to the generative phase of the life cycle shows the plasticity for differentiation of plant cells. In most plant cells the entire plant genome is retained and readily accessible (section 2.3.2).

The plant embryo in the seeds is a very simple image of the mature plant and does not compare with the highly organized animal embryo. After germination plants consist of the root–shoot axis and the cotyledons (*see* section 6.1) and only after the development of the embryo by enlargement of existing cells and cell division can the characteristics of the plant species be seen.

2.6.2 Meristems: the basic units of the plant's developmental plasticity

In the seedling and adult plant, cell division is restricted to the tips of the root–shoot axis. In animals assemblages of cells that are ready to divide are called stem cells. There is a formal resemblance with the plant tissues that can produce meristems (*see* section 6.1). The primary apical meristems of the shoot and the root remain active for longer periods during the plant's developmental cycle and are responsible for the production of new tissues and organs. As new leaf promordia and lateral roots are added over an extended period this is a continuous process. This leads to a modular organization of shoots and roots. Each of the repeated elements (phytomers) consists of nodes, leaf, internodes, and axillary buds in the shoot and of a section of the primary axis in the root (Figure **2.15**).

During the later development of plants additional, non-apical meristems are formed (lateral, intercalary, and the secondary meristems that increase the girth of woody plants). As the primary and secondary meristems are active during the entire developmental cycle, they ensure continuous morphogenesis (for example, branching, and the formation of leaves, flowers, and fruits). This is the reason for stating that the shape of plants is much more flexible than the precisely defined shape of animals.

The plasticity of plant shape is closely linked with its immobile existence at a fixed location. In animals the movement of single cells and the whole organism is a normal part of life but in higher plants there is usually no movement of either single cells or entire organisms.

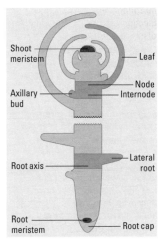

Figure 2.15 Apical meristems and phytomers. The diagram illustrates the modular construction of a plant. Shoot and root meristems repeatedly produce morphological and functional units (phytomers): leaf, node, internode, and axillary bud in the case of the shoot meristem; root axis with lateral roots in the case of the root meristem. As soon as the axillary buds and the lateral roots are fully grown, they form additional phytomers. (Adapted from Lyndon 1990.)

2.6.3 Cell lineages

When we looked at the term *determination* (section 2.5.1) we questioned whether the fate of the differentiated cell also holds true for its offspring. This would result in cell lineages, whose cells have arisen by mitoses and are virtually identical. In animal systems this situation is quite common. The formation of such lineages with fixed cell destiny is important for the development of organs (stem cells; *see* section 2.6.2). However, this can be difficult to prove because of the mobility of animal cells. The movement of

Figure 2.16 Layers of the meristem and the fate of cells. (**a**) In angiosperms, the shoot apical meristem is composed of three cell layers that can be distinguished on the basis of their different location and cell division patterns. The outer (L1) layer is characterized by anticlinal cell divisions (parallel to the surface). The middle (L2) layer and particularly the inner (L3) have both anti- and periclinal divisions (periclinal = perpendicular to the surface). These modes of division result in the tunica and corpus of the shoot (tunica = peripheral cell layers; corpus = central tissue complex; *see* section 6.1). (**b**) This diagram shows that most of the cells in an adult plant originate from divisions in the L3 layer of the shoot apical meristem. The epidermis is formed exclusively by the L1 layer. The L2 layer usually forms multicellular subepidermal layers in the shoot, including the mesophyll of leaves and the cortex of the stem. (Adapted from Poethig 1990.)

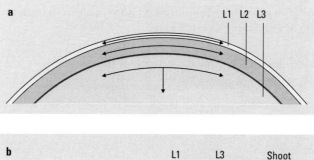

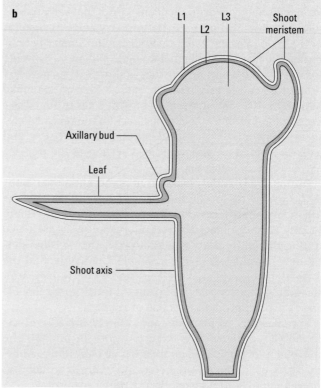

cells to completely new positions within the growing organism is normal during the morphogenesis in the animal embryo. Plant cells, with their cell walls, lack the same mobility, and limited passive movement only occurs after cell division and due to the growth of neighbouring cells. One may conclude from the above that cellular lineages in plants are easily recognizable. Indeed, there is histological proof for these stacked cell groups, whose members have a common developmental origin (i.e. the development of the primary root; *see* section 6.2.1). However, it has been shown that plant cells are generally not influenced by ancestral cells to the same extent as animal cells (*see* section 6.2.5). The apical meristem of the angiosperm shoot consists of three functional layers (L1, L2, L3) which differ in their type of cell divisions and the developmental fate of the cells (Figure **2.16a,b**). If an L1 cell enters, by accident or by experimental design, the L2 layer of the meristem, it will behave like an L2 cell. This suggests that the location of the cell is just as important as its ancestral origin.

2.7 Bibliography

General references

Brownlee, C. and Berger, F. (1995). Extracellular matrix and pattern in plant embryos: On the lookout for developmental information. *Trends in Genetics*, **11**, 344–8.

Burgess, J. (1985). *An introduction to plant cell development.* Cambridge University Press, Cambridge.

Croxdale, J., Smith, J., Yandell, B., and Johnson, J. B. (1992). Stomatal patterning in *Tradescantia*: an evaluatiuon of the cell lineage theory. *Developmental Biology*, **149**, 158–67.

Demura, T. and Fukuda, H. (1994). Novel vascular cell-specific genes whose expression is regulated temporally and spatially during vascular system development. *The Plant Cell*, **6**, 967–81.

French, V. P., Bryant, P. J., and Bryant, S. V. (1976). Pattern regulation in epimorphic fields. *Science*, **193**, 969–81.

Fukuda, H. (1997). Tracheary element differentiation. *The Plant Cell*, **9**, 1157–68.

Ghoshroy, S., Lartey, R., Sheng, J. S., and Citovsky, V. (1997). Transport of proteins and nucleic acids through plasmodesmata. *Annual Review of Plant Physiology and Plant Molecular Biology*, **48**, 25–48.

Goldberg, R. B. (1988). Plants: novel developmental processes. *Science*, **240**, 1460–7.

Goodner, B. and Quatrano, R. S. (1993). *Fucus* embryogenesis: a model to study the establishment of polarity. *The Plant Cell*, **5**, 1471–81.

Hake, S. H. and Char, B. R. (1997). Cell–cell interactions during plant development. *Genes and Development*, **11**, 1087–97.

Hämmerling, J. (1963). Nucleocytoplasmic interactions in *Acetabularia* and other cells. *Annual Review of Plant Physiology*, 14, 65–92.

Jacobs, T. (1997). Why do plant cells divide? *The Plant Cell*, **9**, 1021–9.

Kreuger, M. and Van Holst, G. J. (1996). Arabinogalactan proteins and plant differentiation. *Plant Molecular Biology*, **30**, 1077–86.

Kropf, D. (1997). Induction of polarity in fucoid zygotes. *The Plant Cell*, **9**, 1011–20.

Lyndon, R. F. (1990). *Plant development—the cellular basis*. Unwin Hyman, London.

Marx, J. (1996). Plant biology—Plants, like animals, may make use of peptide signals. *Science*, **273**, 1338–9.

Nultsch, W. (1991). *Allgemeine Botanik*. Thieme, Stuttgart.

Poethig, S. (1989). Genetic mosaics and cell lineage analysis in plants. *Trends in Genetics*, **5**, 273–7.

Poethig, R. S. (1990). Phase change and the regulation of shoot morphogenesis in plants. *Science*, **250**, 923–30.

Reynolds, T. L. (1997). Pollen embryogenesis. *Plant Molecular Biology*, **33**, 1–10.

Russo, V. E. A., Brody, S., Cove, D., and Ottolenghi, S. (1992). *Development—the molecular genetic approach*. Springer-Verlag, Berlin.

Sachs, T. (1991). *Pattern formation in plant tissues*. Cambridge University Press, Cambridge.

Steeves, T. A. and Sussex, I. M. (1989). *Patterns in plant development*. Cambridge University Press, Cambridge.

Sussex, I. M. (1989). Developmental programming of the shoot meristem. *Cell*, **56**, 225–9.

Szymkowiak, E. J. and Sussex, I. M. (1996). What chimeras can tell us about plant development. *Annual Review of Plant Physiology and Plant Molecular Biology*, **47**, 351–76.

Taylor, C. B. (1997). Plant vegetative development: from seed and embryo to shoot and root. *The Plant Cell*, **9**, 981–8.

Taylor, L. P. and Hepler, P. K. (1997). Pollen germination and tube growth. *Annual Review of Plant Physiology and Plant Molecular Biology*, **48**, 461–91.

Waddington, C. H. (1957). *Strategy of the genes*. George Allen & Unwin Ltd., London

Walbot, V. (1985). On the life strategies of plants and animals. *Trends in Genetics*, **1**, 165–8.

Further reading

Becraft, P. W., Stinard, P. S., and McCarty, D. R. (1996). CRINKLY4: A TNFR-like receptor kinase involved in maize epidermal differentiation. *Science*, **273**, 1406–9.

Berger, F., Taylor, A., and Brownlee, C. (1994). Cell fate determination by the cell wall in early *Fucus* development. *Science*, **263**, 1421–3.

Callos, J. D. and Medford, J. I. (1994). Organ positions and pattern formation in the shoot apex. *Plant Journal*, **6**, 1–8.

Knox, J. P. (1992). Cell adhesion, cell separation and plant morphogenesis. *Plant Journal*, **2**, 137–42.

Lucas, W. J. *et al.* (1995). Selective trafficking of Knotted1 homeodomain protein and its mRNA through plasmodesmata. *Science*, **270**, 1980–3.

Medford, J. I. (1992). Vegetative apical meristems. *The Plant Cell*, **4**, 1029–39.

Roberts, K. (1992). Cell signalling: Potential awareness of plants. *Nature*, **360**, 14–15.

Schumaker, K. S. and Dietrich, M. A. (1997). Programmed changes in form during moss development. *The Plant Cell*, **9**, 1099–107.

3 The genetic analysis of developmental processes— methodology

Contents

3.1 **Developmental genes can be recognized by their mutant phenotypes**
3.1.1 Isolation of mutants
3.1.2 Seed or pollen mutagenesis?
3.1.3 Choosing the mutagen
3.1.4 Genetic characterization of mutants
3.1.5 Investigating the effect of the gene
3.1.6 Inferring the normal effect of a developmental gene from its mutant phenotype
3.1.7 When is a developmental gene required?
3.1.8 Does a developmental gene act autonomously?
3.1.9 Are developmental genes also required for regeneration?
3.1.10 The phenotypes of double mutants show the possible interactions between developmental genes

3.2 **From phenotype to gene—how developmental genes can be isolated**
3.2.1 Map-based cloning
3.2.2 RFLP mapping
3.2.3 Chromosome walking
3.2.4 Tracking down the gene
3.2.5 Isolation of molecularly marked genes
3.2.6 From mutagenesis by T-DNA integration to gene isolation
3.2.7 From transposon mutagenesis to isolating a gene
3.2.8 Molecular identification of the gene

3.3 **From the gene back to the phenotype—molecular analyses of the effects of developmental genes**
3.3.1 Investigations of the effects of developmental genes
3.3.2 Occurrence and distribution of the gene product
3.3.3 Regulation of gene expression

3.4 **Prospects**

3.5 **Bibliography**

Preface

According to current understanding, developmental processes are regulated by the effects of interactions between molecules. The identities of the participating molecules are unknown and there are no suitable biological tests that could be used to identify them directly. However, it is possible to identify the genes that encode these molecules or that code for proteins which in turn produce these molecules. That genes can regulate development is demonstrated by the fact that from generation to generation, in both plants and animals, a multicellular adult organism emerges from a single fertilized egg cell. This organism will have an arrangement of differentiated tissues and organs that is typical for its species. This is the basis for using genetic analysis to study development.

The genome of a higher plant might contain 10 000 or more genes that could each affect development in a different way. Many genes code for enzymes or other proteins that participate in the metabolism of every cell. The products of these housekeeping genes are necessary for normal development of the plant but, in contrast to developmental genes, they do not regulate that process. If a developmental gene is missing, its associated development process will have an altered outcome—the plant may have abnormally arranged structures, or some structures could be missing entirely.

Mutant phenotypes are the keys to genetic analyses of development processes. Mutant phenotypes that are caused by inactivation of single genes provide biological test beds for the recognition of developmental genes. Investigations of these mutants enable scientists to hypothesize about the roles of genes in development.

The mutant phenotype can also be used to isolate the gene. Then the molecular structure of the gene can be fully characterized. Molecular data like this can give clues to how the gene products will act, and eventually reveal the basic mechanisms of development.

How well this sort of genetic analysis works depends on the particular plant species. The most suitable ones are referred to as *model systems*. Results from these models are generally valid for other species because flowering plants are closely related. Typical monocotyledonous model systems are maize and rice; for dicotyledonous plants, *Arabidopsis thaliana* is often used. For some specialized investigations, *Antirrhinum*, tomato, and pea plants are used. *Arabidopsis*, a member of the mustard family, is close to being the ideal experimental subject and has become the scientists' favourite for studying the genetics of development in plants (*see* Box **3.1**). Methods described in this chapter will relate to *Arabidopsis* unless stated otherwise.

Box 3.1 *Arabidopsis thaliana*: a model system for genetic analysis of plant development

Arabidopsis thaliana is a wild plant belonging to the mustard family (Brassicaceae). Because it never grows taller than about 30 cm it is easy to grow large numbers in the laboratory under controlled conditions. Under continuous illumination at 25 °C up to 10 000 plants/m² can grow to form ripe seeds within 6 weeks. This means that every year eight generations can be used for experimentation. Its flowers are hermaphrodite, with both male and female sex organs (the stamen and carpel) and normally are self-pollinating. To cross the plants, the stamens from unopened flowers are removed to prevent self-pollination, and then the pollen from another plant is dusted on to the stigma. After selfing, a quarter of the offspring of heterozygous plants will be homozygous. A single flower can produce 30–50 seeds, the whole plant several thousand. These features make the plant ideal for the isolation and genetic characterization of mutants.

The haploid genome of *Arabidopsis* consists of five chromosomes. Several hundred genes have been identified by mutation and mapped on the five linkage groups. Moreover, several hundred molecular markers (RFLPs and RAPDs) have been identified in *Arabidopsis* by comparing different ecotypes. Ecotypes are laboratory lines that have been established from different geographic isolates. Only a few out of the 300 or so known ecotypes are used routinely for mutagenesis and molecular mapping. Examples are *Landsberg erecta* (*Ler*), *Columbia* (*Col*), *Niederzenz* (*Nd*), and *Wassilewskija* (*Ws*). The genome of *Arabidopsis* is relatively small, at approximately 100 Mb, and it has very few dispersed repetitive DNA sequences. This means that *Arabidopsis* is very useful for isolating genes using map-based cloning. *Arabidopsis* can be infected by *Agrobacterium*

tumefaciens and transgenic plants can be produced easily. Such plants are useful for gene expression and regulation studies.

An International *Arabidopsis* Genome Initiative has been set up to sequence the entire nuclear genome of *Arabidopsis* within the next couple of years. As a first step towards this goal, genomic libraries containing large inserts of DNA in yeast artificial chromosomes (YACs), bacterial artificial chromosomes (BACs), or P1 phage vectors are being prepared, to give contiguous stretches (contigs) of genomic DNA. For example, the DNAs of chromosomes 2 and 4, which amount to approximately 40 Mb, have been assembled into 4–6 contigs each. It is estimated that the entire nuclear genome will be available as BAC contigs by 1998, which will make map-based cloning of any gene feasible. Although only about 10 Mb of genomic DNA has been sequenced, the sequencing project is already producing useful information. The average gene density is about one gene per 5 kb, suggesting that the entire *Arabidopsis* genome contains approximately 20 000 genes. Furthermore, about one-third of the coding sequences (ORFs) show similarities to sequences of characterized genes from other organisms deposited in the databases, thus facilitating the functional analysis of *Arabidopsis* genes. To determine the biological role of cloned genes, a reverse genetics approach is necessary. This can be done in *Arabidopsis* by screening large populations of lines that carry independent insertions of T-DNA or transposons for insertions into the gene of interest. Putative mutants are identified as lines that give PCR products with gene-specific and T-DNA or transposon primers. Heterozygous plants are singled out and their progeny are screened for specific mutant phenotypes.

3.1 Developmental genes can be recognized by their mutant phenotypes

The intact nuclear genome (wild-type genotype) ensures normal development leading to a viable fertile plant (wild-type phenotype). Mutations in single genes can lead to arrested development, lethality, altered morphology, or sterility. In many cases, a gene that is essential for normal development, but that does not regulate development, will be affected. To properly identify developmental genes, it is therefore necessary to define the developmental process as precisely as possible, and to isolate mutants with specific differences (mutant phenotypes). To analyse a developmental process completely, all the involved genes should be identified by mutant alleles (*saturation mutagenesis*). To investigate the roles of single genes, the mutagen used should only cause point mutations, so that the mutant phenotypes result from mutations in single genes. Only then is it possible to tell whether a gene participates specifically in the affected development process.

3.1.1 Isolation of mutants

Genes are often inactivated partly or completely by mutations. The mutant alleles are recessive and so only homozygous individuals will have a mutant phenotype. During mutagenesis, different mutation events will occur in a population of cells. Starting from single mutated cells, a mutant line can be established by crossing or selfing (single lines). Relatively few lines will give homozygous mutants with abnormal phenotypes that indicate specific disruptions of development. If the mutations prove to be lethal, or make the plant sterile, sister plants from the same line as these can be used for crossing. Two out of three normal-looking sister plants should be heterozygous and therefore carry the mutated gene.

3.1.2 Seed or pollen mutagenesis?

Mutants can be isolated using either of two procedures: seed mutagenesis or pollen mutagenesis. Seed mutagenesis is less technically demanding and thus more widely used (Figure **3.1**). In this type of mutagenesis, mature seeds with completely developed embryos are used. The embryonic cells are diploid. If a cell mutates, cell division during postembryonic development leads to a heterozygous mutant sector. Only if this sector includes subepidermal cells of the shoot meristem (from which germ cells of the adult plant are produced) are the mutations passed on to

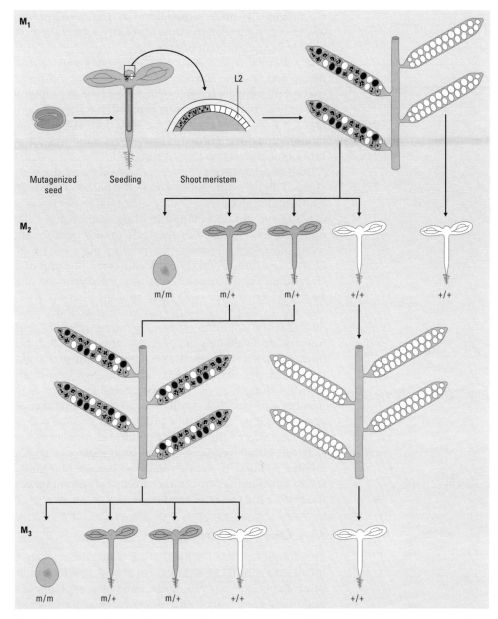

Figure 3.1 Seed mutagenesis. Mature seeds are treated with a mutagen. With a frequency of 1 in 1000–5000 seeds, a gene that is required for normal seedling development will be mutated in one of the two cells that will give rise to the gametes of the mature M_1 plant. A heterozygous mutant sector arises in the L2 layer of the shoot apical meristem and will form the gametes of every other flower. After selfing the flower with mutant gametes, homozygous mutant seedlings (m/m) will occur in the M_2 generation. Two-thirds of normal seedlings which originated from the same pod are heterozygous for the mutation (m/+). Upon selfing the heterozygotes will produce mutant seedlings in the M_3 generation. M_1 is the treated generation; M_2, M_3 are the following generations that are produced by selfing. (According to Jürgens *et al.* 1991.)

the progeny. Flowers included in this sector will produce 25 per cent homozygous mutant progeny, which are recognizable by their mutant phenotype.

In pollen mutagenesis, mature pollen grains are mutated (Figure **3.2**). Pollen consists of three haploid cells—two sperm cells and a vegetative pollen tube cell. The mutation will only be passed to the next generation if the sperm cell that fertilizes the egg cell has the mutation. This leads to a heterozygous mutant that, after selfing, produces 25 per cent homozygous mutant offspring.

Both procedures have their advantages and disadvantages. With seed mutagenesis, in hermaphroditic flowering plants that reproduce by selfing the mutants can be recognized in the next generation by their phenotypes. Pollen mutagenesis requires a further generation. On the other hand, after seed mutation, the resulting plant is a genetic mosaic and only some flowers will carry the mutation. Depending on the number of germ cell precursors, only 12.5 per cent or less will have mutant phenotypes (as opposed to 25 per cent after pollen mutagenesis). These difficulties do not occur when only the progenies of single flowers are evaluated. In contrast, mutated pollen produces F_1 plants with genetically identical cells, so that after selfing 25 per cent of their offspring will be mutant. Another thing that sets the two procedures apart is the number of lines that can be tested. Pollen mutagenesis requires that the F_1 plants be produced by manual pollination. Thus sheer volume of work limits the number of lines that can be tested. Seed mutagenesis does not require manual pollination, so many lines can be tested for mutant traits. Since mutations are rare events, seed mutagenesis tends to be the preferred method.

3.1.3 Choosing the mutagen

The frequency with which a mutation occurs can be increased by using the right kind of mutagen. Which mutagen is to be used depends on the aim of the experiment. If nothing is known about the developmental process to be investigated and many mutants are to be isolated, ethyl methane sulphonate (EMS) is often the mutagen of choice. EMS ethylates guanine, and causes a transition from GC to AT. In this way, point mutations in single genes are generated and there is a high probability that a mutant phenotype will be due to the absence or deficiency of the product of a single gene. Electromagnetic radiation (for example X-rays) is a less efficient mutagen; it causes

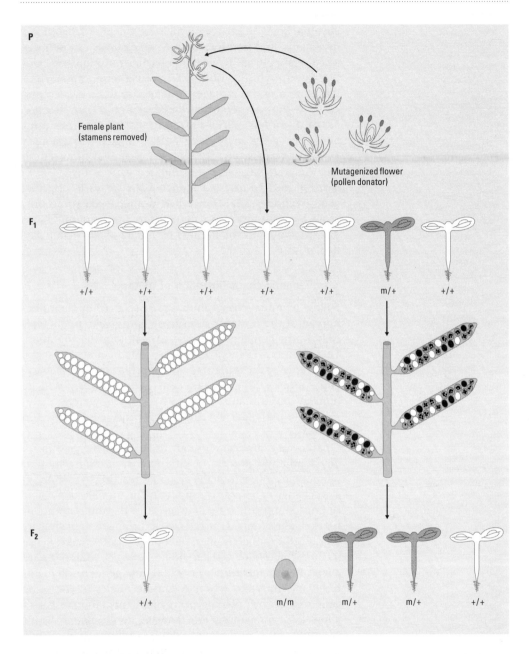

Figure 3.2 Pollen mutagenesis. Flowers with mature pollen are harvested in the parental generation (P). The pollen is mutagenized and used to pollinate plants after removal of their immature stamens. To obtain a sufficiently large number of F_1 plants, as many pollinations as possible have to be carried out. Each F_1 plant represents one mutagenized pollen grain. The F_1 plants are selfed and recessive mutant seedlings (m/m) will become visible in the F_2 progeny. Several thousand F_1 plants have to be studied to identify a mutant allele for a gene of interest.

45

chromosomal breakage, which may lead to deletions, inversions, and translocations. Such changes can be identified relatively easily, using molecular techniques such as Southern hybridization. Electromagnetic radiation is therefore a useful mutagen if genes are to be located precisely within a stretch of cloned genomic DNA (section 3.2.4). Even less efficient is insertional mutagenesis using biological agents such as the T-DNA of *Agrobacterium tumefaciens* or transposons. However, insertional mutagenesis has the advantage that the mutated gene is tagged (*gene tagging*) and the inserted foreign DNA serves to grip that gene (section 3.2.5). How efficient a mutagen is can only really be assessed afterwards by the frequency of mutations or, even better, by the number of mutant alleles of one particular gene.

3.1.4 Genetic characterization of mutants

Following mutagenesis, mutants are selected by their phenotype. If the mutant is sterile or non-viable, phenotypically normal siblings are used to maintain the mutation. These plants will then have to be tested yet again for the occurrence of the mutant phenotype. If a plant is heterozygous, after selfing 25 per cent of its progeny will be homozygous mutants. The raw material for genetic analyses of the particular developmental process is gathered in this way.

Large-scale mutagenesis can yield mutants that affect a developmental process in different ways. These mutant phenotypes reveal how the system is reacting to withdrawal or change of a single component. The number of possible responses appears to be limited, since the mutant phenotypes can be grouped into classes according to their similarities. Further subdivision of phenotype classes makes sense if complementation tests are to prove which mutations are allelic. In general, mutations in different genes result in different phenotypes, whereas mutant alleles of the same gene have similar phenotypes. Complementation tests also allow the mutagenesis efficiency to be estimated. If, for example, on average every gene is represented by five mutant alleles, probably all genes that specifically participate in the developmental process being examined, and that can mutate to form a characteristic phenotype, will have been identified (*saturation*).

Using complementation tests, genes can be defined as functional units called *complementation groups* (Figure **3.3**). If several mutant alleles have similar phenotypes, it can

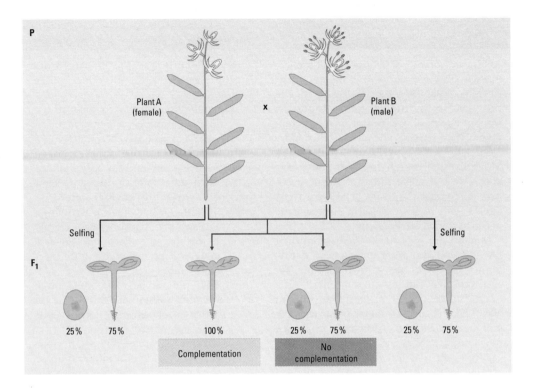

Figure 3.3 Complementation tests. In the case of lethal mutations, only normal plants of lines A and B, which may be either heterozygous or wild type, are available for the crosses. Since only heterozygous plants can be used for the complementation test, all normal plants of lines A and B have to be selfed in order to determine the genetic status at the locus to be tested. The 'female' plants (A) are emasculated and pollinated with the 'male' plant (B). If all the offspring of the crossing of A and B are normal (i.e. A and B complement each other), A and B carry mutations in different genes. If normal as well as mutant offspring are produced, A and B do not complement each other, and the mutant loci in A and B are alleles of the same gene.

safely be assumed that this phenotype is specific, reflecting a lack of, or reduction in, the gene's normal effect. In this sense, the mutant phenotype is like a 'fingerprint' of the gene. Another characteristic value of the gene is its position on the genetic map: the mutant allele of the gene is mapped against previously mapped mutants, called markers, by meiotic recombination.

3.1.5 Investigating the effect of the gene

Mutants are isolated because of their phenotype. To assess how a gene acts in development, several mutant alleles must be isolated (Box **3.2**). If all the mutant alleles show the same phenotype, it would indicate that a critical amount of

Box 3.2 Genetic terminology

Alleles are variants of genes with differences in their DNA sequences. In a mutant allele, the biological effect of a gene is altered, and this appears as the mutant phenotype. Mutant alleles of a gene can have different phenotypes: 'strong' alleles will differ more from the wild type than 'weak' ones. Molecular analysis often shows that strong alleles do not make functional gene products. In comparison to the wild type, weak alleles tend to have slightly altered gene products which are less active.

Mutants are isolated on the basis of their phenotypes. Saturation mutagenesis is performed in such a way as to ensure that for each gene that can mutate to a certain phenotype there will be at least one mutant allele. Statistically, saturation mutagenesis is said to be reached when, on average, at least five mutant alleles have been isolated for each gene. It is assumed that mutations are rare events and that different genes mutate at similar frequencies. Whether mutants with similar phenotypes are alleles can be investigated using complementation tests: if two mutants do not complement each other, they are alleles of the same gene (complementation group). Complementation between alleles is rare, and most of the time incomplete.

Usually several genes that can mutate to form related phenotypes participate in developmental processes. The phenotypes of double mutants will show whether these genes act independently, or if they take part in the development process:

1. The phenotype of a double mutant is the sum of the phenotypes of the single mutants. In this case the two genes act independently and the failure of one gene does not affect the effects of the other gene.

2. The double mutant looks similar to one of the single mutants, which is then said to be *epistatic*, whereas the phenotype of the other single mutant is unrecognizable in the double mutant. Epistasy indicates that both genes interact during development. There are two kinds of epistasy:

    (a) The phenotypes of the two single mutants are opposites (*see* Figure **3.5**). In this case, the effect of the epistatic gene will be negatively influenced by the other gene.

    (b) The phenotypes of both single mutants are similar. In this case the activity of the epistatic gene is a prerequisite for the other gene to become effective. Genetic findings cannot provide any information on the nature or mechanism of the interaction. Only molecular analysis can show if the product of one gene can directly regulate the expression of the other gene. This regulation might be by transcriptional control, or if the effect is indirect, one gene product might promote or inhibit the activity of the other gene product.

3. The double mutant might have a new phenotype that is neither due to addition nor epistasy of the phenotypes of the single mutants. This rare effect can occur if both genes act in concert during development, so that inactivation of only one gene perturbs but does not completely inhibit the process.

gene activity is needed for the gene to express its normal effect. Alternatively, alleles can differ in the severity of their phenotypic deviations from the wild type. In this case the gene's effect is determined directly by the amount of gene activity. The biggest deviation from the wild-type phenotype would probably indicate that the gene is completely inactive, while intermediate phenotypes indicate that there is still some residual gene activity. Either way, the phenotype that results from complete inactivity can provide clues as to whether the gene plays a specific role in development.

3.1.6 Inferring the normal effect of a developmental gene from its mutant phenotype

The mutant phenotype of a developmental gene results from disruption of the normal course of development, due to the lack of a single component. Precisely what part the affected gene plays in normal development can then be inferred by careful investigations of the development of mutant individuals. Take, for instance, a mutant seedling without any roots. If the early stages of embryo development are studied, the stage when the development of roots in abnormal plants diverges from that of normal roots can be ascertained (*see* Figure **6.1**). In this way, the time when the gene product is first needed for normal root development can be estimated. Sometimes it is possible to see the first manifestation of the mutant gene; for instance, it may be an abnormal cell division in the embryo. However, this does not help elucidate the role of the normal gene product in cell division, but at least the mutant phenotype can be more precisely defined at the cellular level, rather than at the level of the organism, so we are a step closer to finding the actual effect of the gene product.

3.1.7 When is a developmental gene required?

Investigations with mutant individuals make it possible to narrow down the place and time of the gene's effect. However, it is usually impossible to say whether the gene's effect on a developmental process is short lived and only happens once, or whether the gene can affect the process continuously. In a few cases, temperature-sensitive mutants of the developmental gene can be isolated; they can help to solve this problem.

Temperature-sensitive alleles have the following features: they allow normal development at permissive temperatures (usually low temperatures) but not at the restrictive temperature, which is usually higher. Wild-type plants are not affected this way. Deliberately changing the temperature as a mutant plant develops can show whether the mutated gene is used at different times during development, or just once. In this way it can be seen whether a gene that is needed during embryonic development (the absence of which would lead to abnormal seedlings, unable to develop any further) is also necessary for later, postembryonic, development. If no temperature-sensitive allele is available, we can see whether an early acting developmental gene is used later in development by studying the regeneration of shoots and roots from mutant calli.

3.1.8 Does a developmental gene act autonomously?

Genetic mosaics and chimeras can also help in investigating time-dependent gene effects. Furthermore, these techniques are used to determine whether a gene acts autonomously (independently of the cell). Genetic mosaics are made by changing the genotype of single cells during development. After a few rounds of cell division these altered cells form sectors (Figure **3.4**). Such mutant sectors can be induced by mutagenesis of heterozygous plants (cf. seed mutagenesis, section 3.1.2). If the sector has a mutant phenotype, this suggests that the gene acts autonomously. However, if the sector has a normal phenotype, the gene obviously is not required in the cells of the sector. This can be explained in different ways: the gene might not be needed during the developmental stages that were investigated, or the normal neighbouring cells can compensate for any lack of the gene product in the sector.

Chimeras are also combinations of mutant and wild-type tissue in a single plant. They differ from genetic mosaics in that they are produced when complementary parts of two genetically different individuals grow together to produce a single whole. For example, a mutant seedling that was missing roots could be grafted on to the hypocotyl of a wild-type seedling. If both parts grow together, such a

Figure 3.4 Genetic mosaics as test systems for autonomous gene effects. Seeds with heterozygous embryos (*gl1*/+) are mutagenized. According to a certain probability, the intact wild-type allele in one or the other cell becomes inactivated and the respective cell will form a mutant sector, provided that the affected gene acts autonomously. A gene required for the formation of trichomes, *GL1*, is used here as an example (*see* section 6.2.7).

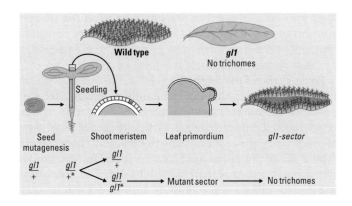

chimera could be used to see whether the shoot meristem of the mutant part develops normally, eventually producing fertile flowers.

3.1.9 Are developmental genes also required for regeneration?

Plants can regenerate missing parts. If a seedling is cut through at the hypocotyl, the upper part of the seedling will grow new roots under favourable conditions. If, during embryonic development, a developmental gene fails, and as a result the seedling has no root, we can investigate whether the mutant seedling can still establish a root if given the necessary inductive conditions. If it can, we can conclude that the particular developmental gene is not required to produce the root. The developmental gene seems to establish whether a root can form in the lower region of the embryo. This means that root formation during embryogenesis depends on different factors to those necessary for regeneration.

3.1.10 The phenotypes of double mutants show the possible interactions between developmental genes

A defect in a single gene may have a strong effect on a developmental process. For example, a completely different flower might be produced. Even though the mutant phenotype reflects the effect of the gene, it is by no means certain that only this gene can create this phenotypic appearance. For instance, mutations in several genes can lead to changes in flower structure, and this raises the question as to how the genes interact with one another (*see* section 6.3.2). This question can be tackled by a phenotypic analysis of double mutants, as long as the single mutants can be distinguished by their phenotypes (Figure **3.5**). If the 'sum' of the phenotypes of the single mutants equals that of the double mutant, it can be assumed that the affected genes act independently in the process. Alternatively, the double mutant may only have the phenotype of one of the single mutants. This is known as *epistasy*, meaning that one gene influences the effects of another (Box **3.2**). This interaction between two genes does not necessarily mean that the product of the first gene regulates expression of the second gene, in the same way that the effect of a downstream gene in a process can be dependent on an upstream one. Whether the interaction is direct or indirect can only be determined by molecular

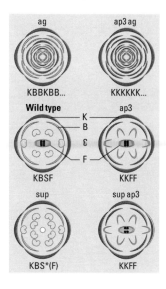

Figure 3.5 Phenotypic analysis of double mutants. The flower mutants *ag*, *ap3*, and *sup* are used to illustrate how the generation of double mutants helps to determine whether the genes interact with one another. Wild-type flowers consist of four concentric whorls. In *Arabidopsis* there are four sepals (K), four petals (B), six stamens (S), and two carpels (F). In the *ag* mutant, stamens are transformed into petals (S › B) and, instead of forming carpels, the flower programme is repeated several times. In flowers of *ap3* mutants petals are transformed into sepals (B › K) and stamens into carpels (S › F). In *sup* mutants the number of stamens increases at the expense of carpels (*see* section 6.3.2). The phenotype of the *ap3 ag* double mutant is the sum of the phenotypes of the two single mutants. It follows from this that *ag* and *ap3* act independently of one another. The *sup ap3* double mutant shows the phenotype of *ap3*, i.e. *ap3* is epistatic to *sup*. Accordingly, one gene (*SUP*) may down-regulate the other (*AP3*). Alternatively, *AP3* may be required to activate *SUP* (*see* section 6.3.2).

51

analyses of the participating genes and their products. Nevertheless, phenotypic investigations of double mutants allow statements to be made about the network of gene effects, and thus precise hypotheses to be proposed about the biological roles of single genes during development.

3.2 From phenotype to gene—how developmental genes can be isolated

Two properties of identified developmental genes are immediately apparent: the mutant phenotype and its position in the genome. This information is sufficient to isolate a gene if one of the following strategies is used:

1. *Map-based cloning.* The gene is mapped against molecular markers (DNA polymorphisms) and chromosome walking is started from the marker nearest to the gene. This method is best employed in plants with small and compact nuclear genomes with only a few, widely distributed repetitive sequences.

2. *Gene tagging.* Foreign DNA sequences, for example T-DNA or transposons, that are inserted into a gene are used as a probe to isolate flanking plant DNA sequences. Successful insertion mutagenesis is a prerequisite for this method to work. Not every gene can be mutated using this method, and in general it requires significant effort.

3.2.1 Map-based cloning

Map-based cloning is based on the classical principle of gene mapping using meiotic recombination: as recombination frequency between a genetic marker and the gene to be mapped decreases, so does the genetic distance between them (measured in centiMorgans, cM). However, instead of morphological markers, DNA polymorphisms are used as markers (Box **3.3**). If possible, very closely linked markers are identified that will be useful as starting points for chromosome walking. How much of a distance needs to be bridged to reach the gene depends on the genome size and the frequency of recombination in the investigated region. In *Arabidopsis*, 1 cM is, on average, about 200 kb since the genome size is approximately 100 000 kb with a genetic length of roughly 500 cM. The actual distance that equates to 1 cM depends on the genome region, and might be anywhere from 100 to 1000 kb.

Box 3.3 Molecular markers (DNA polymorphisms)

Molecular markers are defined as DNA sequence differences that occur at homologous positions in the genomes of different individuals belonging to the same species. These *DNA polymorphisms* are phenotypically neutral alleles that can be identified using techniques in molecular biology, such as Southern hybridization and PCR amplification.

In *Arabidopsis*, genomic differences between lines with different geographic origins (ecotypes) are used as markers. The markers used most often are:

1. RFLP (*restriction fragment-length polymorphism*). The genomic DNAs of two ecotypes are digested with a restriction enzyme, separated by gel electrophoresis, then blotted on to a membrane and hybridized with a radioactively labelled probe. If, within the genomic area that the probe can recognize, the two ecotypes have differently positioned restriction enzyme recognition sequences, the banding pattern of the two ecotypes will be different on the autoradiograph. The polymorphism can occur either because in one ecotype the recognition sequence for the restriction enzyme is lost through mutation (one instead of two bands) or because the distance between two neighbouring recognition sites has increased after an insertion ('higher' and 'lower' bands). RFLPs behave like co-dominant alleles of a gene. Their usefulness depends on suitable restriction enzymes for the polymorphism being found.

2. RAPD (*randomly amplified polymorphic DNA*). A single primer (a short oligonucleotide with a randomly chosen sequence) is used to amplify genomic DNA by PCR. The DNA fragments that result from this are separated electrophoretically. This technique relies on the fact that for a fragment to be amplified, the primer must bind to the forward DNA sequence at some point and also to the reverse DNA sequence somewhere downstream of this. The DNA polymorphisms are dominant/recessive if the necessary complementary primer sites only occur in one ecotype, or co-dominant if the distance between them is different. Many different RAPDs can be identified quickly using many different primers.

3. CAPS (*cleaved amplified polymorphic stretch*). Specific portions of the genomes of two ecotypes are amplified by PCR using two sequence-specific primers. The amplified fragments are cut with a restriction enzyme and separated by gel electrophoresis. If the amplified region of one ecotype contains the recognition site of the restriction enzyme but the other does not, the polymorphism will be seen on the gel as two bands instead of one. CAPS are co-dominant markers that are usually obtained by subcloning RFLP markers. They are especially useful for genotyping plants when large mapping populations are to be investigated.

4. SSLP (*simple sequence length polymorphism*). Many genomes contain tandem arrays of multiple copies of two or three nucleotides (simple sequences) at many positions. The number of copies can vary between individuals or lines. This length polymorphism can be revealed by PCR amplification and gel electrophoresis if unique flanking sequences are used as specific PCR primers. A set of SSLP markers covering the *Arabidopsis* genome have been developed for molecular mapping of phenotypically identified genes.

3.2.2 RFLP mapping

Arabidopsis lines used in the laboratory that have been grown from seeds with different origins (ecotypes) differ by approximately 1 per cent of the nucleotide sequence of their genome. These changes can lead to a restriction enzyme site being intact in one ecotype (for example *Landsberg erecta, Ler*) but not in another (for example *Niederzenz, Nd*). Genomic DNA regions are investigated in the following way for polymorphisms: genomic DNA from both ecotypes is digested with the same restriction enzyme, and the resulting DNA fragments are separated by gel elec-trophoresis. The gel is blotted on to a membrane filter, and the DNA fragments on the filter are hybridized to a radioactively labelled genomic DNA probe. X-ray film is then exposed to the filter, and DNA fragments of different lengths to which the probe hybridized are revealed. This technique is known as *restriction fragment-length polymor-phism (RFLP)*, and the corresponding sites in the genome are called *RFLP markers*.

To map a gene using RFLPs, a segregating population must be generated (Figure **3.6**). If, for example, a mutant allele has been induced in the *Ler* ecotype, then the het-erozygous mutant (or if it is not lethal, the homozygous mutant) plants are crossed with the wild type of the *Nd* ecotype. F_1 plants that are heterozygous for this gene are also heterozygous for all *Ler/Nd* polymorphisms. During meiosis, recombination events occur that separate the gene from the previously linked RFLP markers, and genetically distinct F_2 individuals are produced. So that there is enough DNA for an RFLP analysis, the F_2 plants are selfed and the resulting F_3 families analysed by RFLP. Each family represents the genotype of the F_2 parent. To find a tightly linked DNA polymorphism, 50 F_3 families are sufficient. Since meiotic recombination occurs in both F_1 female and male meioses, the resulting mapping accuracy is 1 per cent (a distance of 1 cM). For mapping, two kinds of RFLP markers can be used: ones that represent certain regions of the genome, or ones from the region in which the gene has been found by use of morphological markers. The aim of this mapping exercise is to find RFLP markers that are 1 cM or less away from the gene.

3.2.3 Chromosome walking

A closely linked DNA polymorphism is identified for use as a starting clone for chromosome walking, then the really hard work begins. Genomic DNA clones must be identified

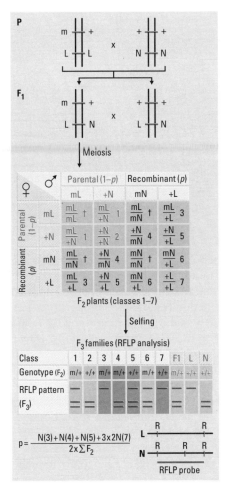

Figure 3.6 RFLP mapping. A developmental gene is mapped against an RFLP marker that is polymorphic for the ecotypes *Ler* (L) and *Nd* (N). A mutation in the developmental gene was induced in the *Ler* ecotype and results in lethality of the homozygous mutant (m/m). Heterozygous (m/+) *Ler* plants are crossed with wild-type plants (+/+) of the *Nd* ecotype. In the F$_1$, 50 per cent of the plants are m/+ (L/N). During meiosis of the m/+ plants, female and male gametes are formed which are either parental (m/L or +/N) or recombinant (+/L or m/N). Random mating results in seven genotypical classes (1–7) that do survive. F$_3$ families are produced from each F$_2$ plant by selfing. This allows determination of the genetic status at the locus to be mapped, i.e. whether the F$_2$ plants were homo- or heterozygous (+/+ or m/+). The pooled DNA of the F$_3$ family provides enough material to genotype each F$_2$ plant (L/L, N/N, or L/N) from which the F$_3$ family was generated. Genotyping is carried out by Southern hybridization with RFLP probes or by using any other molecular marker system (*see* Box **3.3**). The genetic distance between the RFLP marker and the locus to be mapped is calculated by the meiotic recombination frequency (p) of the F$_1$ plants. Since the recombinant class (6) cannot be discriminated from the parental class (1), class 7 is multiplied by 3. Class 7 is also multiplied by 2, because the two homologous chromosomes are both recombinant.

and sorted relative to one another. These clones must be located between the starting point and the gene to be isolated (Figure **3.7**). From a genomic DNA library, DNA clones that overlap (that is, those with common genomic sequences) can be isolated by hybridization with the initial clone. The new DNA clones or fragments are also hybridized to the DNA library to identify further overlapping DNA clones that contain neighbouring parts of the genome. If this procedure is repeated several times, chromosome walking in opposite directions results in a contiguous genomic area of overlapping DNA clones, all based around the initial clone. Now this cloned area of the genome needs to be oriented in relation to the gene. The clones that lie at each end are tested for DNA polymorphisms that can be used for mapping in the segregating

Figure 3.7 Chromosomal walking. Two flanking RFLP markers (R1 and R2) are the starting points for the walk to the gene to be isolated (m). The RFLP probes are hybridized to yeast colonies containing yeast artificial chromosomes (YACs) with genomic DNA from *Arabidopsis*. The ends of the isolated YAC clones are amplified by PCR and, if polymorphic, they are mapped using a segregating population with at least 2500 meiotic events. The ends of YAC clones that map closest to the gene are then used to isolate new YAC clones that are again are mapped against the gene and the previously isolated YAC clones. When YAC clones are available that span the genes, these are used to identify cosmid clones. The cosmid clones are mapped with respect to the gene and are assembled to a contig. In the example shown here there is one cosmid clone that spans the gene.

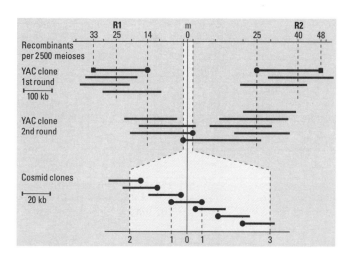

population. The end clone which is closer to the gene of interest than is the initial clone detect fewer recombinants than the clone at the opposite end. This is because recombination frequency is proportional to the genetic distance.

The speed of chromosome walking depends on two factors: the theoretical resolution during fine mapping and the size of the cloned fragments. The term *theoretical resolution* refers to how the size of the segregating population limits how markers can be mapped to one another and to the gene. For example, the real distance cannot be determined if in 50 F_3 families no recombination between the initial clone and the gene can be found. However, in a population ten times larger, six recombinations might be found, and the distance would be approximately 0.6 cM. This simple example demonstrates that a mapping population should be as large as possible to enable small distances to be determined. However, to keep the molecular genotyping work to a minimum, as small a number of F_2 plants as possible (that is, the ones with the most conclusive genotypes) is selected. Only F_3 families produced by these plants are used for fine mapping. For the preselection process, RFLP markers that flank the gene by a few centimorgans are converted into PCR markers (CAPS, Box **3.3**). Using this technique, 1000 or more F_2 plants can be genotyped by PCR analysis. This can identify the few plants that are recombinant in this region. In 1000 F_2 plants, the theoretical solution would be 0.05 cM. This equates to 10 kb, on average. For fine mapping, though, only 50 F_3 families need be analysed if the size of the preselected region is 5 cM.

For efficient chromosome walking, large DNA clones have to be used. These are useful, because with just a few steps, large distances can be covered. This concept eventually led to the creation of YAC libraries of genomic DNA from *Arabidopsis*, amongst other species. Each DNA fragment, which is up to 600 kb long, is integrated into a yeast artificial chromosome (YAC). These are extra chromosomes that are introduced into and maintained in yeast cells. A library of 2500 yeast colonies, each possessing 150 kb YACs on average, will contain 3–4 equivalents of the genomic DNA of *Arabidopsis*. With chromosome walking, the relevant YAC clones can be identified using colony hybridization with the initial clone. The ends of the genomic DNA inserts are amplified and used as probes to identify more YAC clones. In this way a contiguous stretch of several hundred kilobases can be covered rapidly. The real distance can also be measured: using pulsed field gel electrophoresis the size of the genomic DNA insert in a YAC clone can be estimated. The ends of each identified YAC clone can be hybridized with the entire set of YAC clones from the genomic region of interest, putting all the YAC clones in order. Mapping the polymorphic ends in the high-resolution mapping population helps in estimating the recombination frequency and correlates it to the real distance. In this way a physical map of the genomic region that contains the gene (Figure **3.7**) begins to emerge. The chromosome walking process is finished when both ends of a YAC clone, or the neighbouring ends of overlapping YAC clones, recognize different recombinants. Now the gene must be tracked down in this part of the genome.

3.2.4 Tracking down the gene

To locate the gene more accurately, the part of the YAC clone identified as containing the gene is used to screen a genomic cosmid library. Each clone in this library will only contain 20–40 kb of genomic DNA. The positive cosmid clones are then ordered with respect to one another. The clones that recognize DNA polymorphisms are mapped to the segregating population (Figure **3.7**). In an ideal situation, only a few cosmids will be left, which together should contain the gene. At this point, the recombination mapping has reached its practical limit. To locate the gene more precisely would take a great deal more effort.

For the final steps in locating the gene, two methods are used: complementation of a mutant allele with wild-type DNA in transgenic plants and the location of allele-specific DNA polymorphisms. For complementation, cosmid clones are inserted into the genomes of (heterozygous) mutant plants using T-DNA transformation (*see* section 7.2.3). If the homozygous mutant offspring of the transgenic plants develop normally, all sequences necessary for correct expression of the gene are contained in the inserted DNA. Cosmids containing only part of the gene cannot suppress the mutant phenotype. In this way the gene's location can be narrowed down to a particular cosmid or to the overlapping region of two cosmids. Alternatively, or additionally, the gene can be more precisely located by mapping alleles that have had chromosomal mutations induced by X- or gamma-radiation. In the area of the gene, chromosomal mutations would alter the restriction pattern in comparison to the starting lines (the wild-type ecotype). This allele-specific DNA polymorphism allows the gene to be tracked down to be within just a few kilobases of DNA in the cloned genomic region. After this, the gene can be clearly identified (section 3.2.8).

3.2.5 Isolation of molecularly marked genes

A direct way to isolate genes is to mark the gene molecularly with cloned foreign DNA (*gene tagging*) which is then used as a probe to isolate flanking plant DNA. Variants of this strategy differ in the way that a gene to be isolated is tagged: T-DNA from *Agrobacterium* is integrated stably, whereas endogenous or foreign transposons are integrated unstably. To clone developmental genes this way, alleles resulting from insertion mutagenesis must be isolated. Not every gene can be cloned this way, because the integration of foreign DNA is not by chance alone. On average, insertion mutagenesis is about 50 times less efficient than chemical mutation using EMS, but if a gene is tagged, the flanking genomic DNA can be cloned in several ways. For example, genomic DNA from the insertion mutant is digested with a restriction enzyme that cuts outside of the inserted DNA sequence and religated. Specific PCR primers that anneal with the ends of the inserted DNA and face outward enable the short stretch of flanking plant DNA to be amplified by PCR. This technique is called inverse PCR. Alternatively, following religation, the circular DNA containing the inserted DNA and flanking plant DNA can be transformed into *Escherichia coli* for propagation as a plasmid. This technique, which is called plasmid rescue, requires the presence of an origin of replication and a selectable marker, such as resistance to an antibiotic. Gene tagging thus circumvents the time-consuming chromosome walking. A general problem with isolating molecularly marked genes is that it needs to be shown that the insertion of foreign DNA, and not just a random mutation, is responsible for the mutant phenotype.

3.2.6 From mutagenesis by T-DNA integration to gene isolation

T-DNA introduced by *Agrobacterium* into the plant cell integrates into the nuclear genome without any noticeable sequence specificity. For mutagenesis, a modified T-DNA is used that, amongst other things, has a marker gene that will make the transformed plant resistant to antibiotics such as hygromycin or kanamycin. Mutants are then isolated from the offspring of the transformed resistant plants according to their phenotypes. It is assumed that the integration of the T-DNA into a gene hinders the production of an active gene product, resulting in a mutant phenotype, but this is not always the case. Therefore a segregation analysis has to ensure that in a larger segregating population antibiotic resistance is always associated with the mutant phenotype (Figure **3.8**). In random spot checks of about 100 *Arabidopsis* mutants produced by T-DNA integration, segregation analysis showed that only a third of the

Figure 3.8 Segregation analysis of T-DNA insertions. A T-DNA carrying a gene for kanamycin resistance (Kan^R), is used to generate a mutant (m). Segregation analysis of mutant phenotype and kanamycin resistance has to verify whether the mutant is indeed caused by the insertion of the T-DNA. Heterozygous KanR/+ plants are selfed. During meiosis parental (mKan^R or ++) and recombinant (m+ or +KanR) gametes are formed. Upon random mating they produce seven genotypic classes of viable offspring (1–7). Selection for kanamycin resistance is applied and the surviving plants are tested for the production of m/m offspring. If all of the many kanamycin-resistant plants tested produce m/m offspring, the T-DNA insertion is probably the causal agent of the mutant phenotype. However, if KanR plants can be identified that do not produce mutant progeny (for example class 5), this means that the T-DNA insertion can be separated from the mutant site by recombination and that the T-DNA cannot be used to isolate the mutant gene.

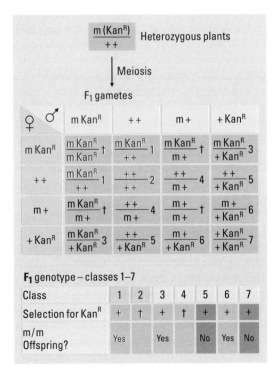

mutants had the desired association between mutant phenotype and antibiotic resistance. In the majority of mutants the resistance was separated from the phenotype by recombination.

Using various methods, about 30 000 lines of *Arabidopsis* have been established that contain modified T-DNA stably integrated into their genomes. In early studies, leaf discs or root pieces were infected with agrobacteria and resistant calli were grown from the successfully infected parts. From these, transgenic plants were regenerated. Other methods, in which tissue culture is not necessary, make use of the ability of agrobacteria to infect germinating seeds, or wounded or intact plants. *In planta* transformation involves either wounding a plant by cutting off its shoot at the stem base and infecting the wound with agrobacteria, or infecting an intact plant using vacuum infiltration. In this transformation without regeneration, treated plants produce very few offspring with successfully integrated T-DNA. Obviously the T-DNA is integrated into the genomes of single cells very late on in development.

3.2.7 From transposon mutagenesis to isolating a gene

Mutagenesis using transposons offers two advantages over T-DNA mutagenesis. Transposons that are already present in the genome can be used for further mutagenesis, whereas T-DNA, once it is successfully integrated into the genome, stays at the same location. Any further mutagenesis of T-DNA-transformed plants would require that they are re-transformed with T-DNA. The second advantage also stems from the mobility of the transposons. If a transposon causes a mutant phenotype by being inserted into a gene, the resulting mutant allele will be unstable because the transposon might leave the gene later on. Such a reversion to the wild-type phenotype is good evidence that the mutant phenotype was indeed the result of a transposon insertion.

During transposon mutagenesis it is necessary to differentiate between plant species that naturally possess active transposons (*endogenous transposons*: for example *Zea mays* and *Antirrhinum*) and those species (for example tomato and *Arabidopsis*) into which foreign (heterologous) transposons must be introduced using T-DNA. For mutagenesis with heterologous transposons, *two-component* systems have been developed (for example Ac/Ds from *Zea mays*). Such systems have several advantages. One of the components is a defective transposon, such as the Ds element, which does not transpose by itself, but nevertheless can be mobilized by transposase. The second element, which might, for example, be derived from the Ac element, produces transposase under the control of a strong promoter. This second element is defective and cannot be mobilized by transposase because it lacks the necessary recognition sequence. The constructs also contain markers for selection which make it possible to recognize plants with new Ds element insertions (Figure 3.9).

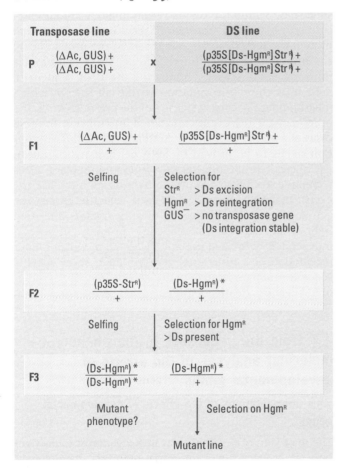

Figure 3.9 Mutagenesis with heterologous transposons. Two transgenic *Arabidopsis* lines are used. The transposase line (ΔAc, Gus) carries the transposase gene (ΔAc) from the Ac element of maize. The transposase gene is stably inserted into the genome next to the β-glucoronidase gene (GUS) which serves as a readily detectable marker for the presence of the transposase gene. Both genes are driven by constitutive promoters. The Ds line (p35S[Ds-Hgm^R]Str^R) carries the non-autonomous Ds element from maize, which is marked by a gene coding for hygromycin resistance (Ds-Hgm^R). The thus modified Ds element is, in turn, inserted into a streptomycin-resistance gene driven by the 35S promoter of cauliflower mosaic virus. The insertion of the Ds element inactivates the Str^R gene which otherwise would be constitutively active. The F_1 progeny of a cross of the two lines contain both transgenes, and the non-autonomous Ds element can be mobilized by the transposase. Selfing of F_1 plants generates the desired F_2 offspring: (1) the plants are resistant to streptomycin because the Ds element has been transposed and thereby the Str^R gene is reactivated; (2) the plants do not possess the transposase gene and hence have no GUS activity; and (3) the plants are resistant to hygromycin, which indicates that the Ds element has been transposed to a new site within the genome. Selfing of these F_2 plants allows the isolation of mutants that are homozygous for the Ds element. To verify that the Ds element is the causal agent of mutation, the transposase gene is crossed into the mutant and revertants are selected. Reversions to wild type can be expected if the Ds element excises 'precisely' and therefore restores the original gene function.

3.2.8 Molecular identification of the gene

It has to be proved that the isolated genomic DNA sequences contain the developmental gene. First, clones are isolated from a cDNA library which may represent the mRNA of the gene. By comparing these cDNA clones with the matching genomic DNA the transcribed region of the gene can be mapped. Finally, the genomic DNA sequences of the wild-type and mutant alleles of the gene are compared. EMS will cause point mutations. A change of just a single base within the coding region might lead to premature termination of translation, resulting in a truncated protein product, or to an amino acid in the protein being substituted for another, unrelated one. These point mutations are easily seen by comparing the DNA sequences. Seldom is the point mutation within an intron, which might hinder the correct splicing of the pre-mRNA. On the other hand, T-DNA is often inserted into introns. In these cases, the mRNA could not accumulate stably; this assumption needs to be proved by RNA gel blot hybridization (Northern hybridization) of RNA from mutant plants (This technique is similar to Southern hybridization (cf. Box **3.3**) except that RNA, instead of DNA, is separated by agarose gel electrophoresis, blotted on to a membrane, and then hybridized with a radiolabelled probe.). If several independently isolated mutant alleles, all of which hinder production of a functional gene product, show specific sequence changes relative to the starting line, the developmental gene is clearly identified.

3.3 From the gene back to the phenotype—molecular analyses of the effects of developmental genes

3.3.1 Investigations of the effects of developmental genes

The main aim of the molecular investigation is to discover how the isolated developmental gene exerts its biological effect. In some cases, the amino acid sequence derived from the cDNA may point to the function of the gene product; for example, known structural motifs of transcriptional factors or components of the signal transduction chain could be identified. In these cases, the route to further characterization of the gene products would be obvious. It is not as easy to proceed if the deduced amino acid sequence of the gene product provides no clues about its function. In any case, it is useful to know when and where during development the gene product is produced, how it is distributed within the cell, and whether it interacts with other proteins with known functions.

3.3.2 Occurrence and distribution of the gene product

Northern hybridization and *in situ* hybridization of tissue sections can show when and where a developmental gene is expressed at the level of transcripts (Box **3.4**). After specific antibodies have been raised against the gene product, the distribution of the encoded protein can be investigated during development as well as in individual cells, using indirect immunostaining of tissue sections. If the deduced amino acid sequence yields no clues whatever to the likely function of the gene product, it is nevertheless useful to know if the protein is located within the nucleus, cytoplasm, cell membrane, or elsewhere. Information on the gene's expression is compared to the known effects of the gene, which have already been deduced from studies of the mutant phenotype. If, for example, the mRNA is ubiquitous throughout the plant, and the lack of gene function only affects the root primordium in the embryo, it is possible that expression of the gene is regulated by translation. Alternatively, a ubiquitously occurring gene product could be activated locally if it is processed by, or interacts locally with, a different molecule.

3.3.3 Regulation of gene expression

If transcription of a developmental gene is limited to a certain time or location, it is possible to investigate whether development could be altered if the gene is expressed abnormally, for instance by fusion of the cDNA with a constantly active promoter (Box **3.4**). If this is the case, the biological effect of the gene would largely correspond with the regulation of its expression. Transgenic plants containing different parts of the promoter of the developmental gene fused to a reporter gene can then be used to track the *cis*-regulatory elements of the gene (*see* section 4.1.4). These same constructs can be used for cross-breeding into mutant plants in order to find transfactors that regulate the developmental gene. Thus we can test whether a locally expressed developmental gene that codes for a transcription factor influences its own expression; autoregulation such as this has been found in genes for flower development (*see* section 6.3.2). Mutations in other genes that are part of the same developmental process can be examined

---

Box 3.4 Investigating the expression of developmental genes

The effects of developmental genes are limited to specific times or locations. How their effects depend on their expression can be investigated in the tissue using several different methods:

1. *In situ* hybridization using a gene-specific probe makes the distribution of the gene transcript visible in tissue sections. By transcribing a cDNA *in vitro*, one can produce antisense RNAs that bind to the gene-specific mRNA in the tissue. Radioactive probes can be detected by coating the sections with a photographic emulsion, and non-radioactive, digoxygenin-labelled probes can be detected by specific antibodies that are conjugated with an enzyme that can catalyse a colour reaction.

2. Indirect immunofluorescence or indirect immunostaining can be used to detect the protein product of a gene. The tissue is incubated with the specific (primary) antibody against the gene product and then with a secondary antibody that binds to the primary antibody. The secondary antibody is conjugated with a fluorescent molecule (colour substance) or with an enzyme that catalyses a colour reaction.

3. In a transgenic plant, a reporter gene can be fused to a developmental gene to aid investigation of that gene. The reporter gene's product catalyses a colour reaction, showing exactly where the fused developmental gene is expressed. For example, the *cis*-active region of the developmental gene is fused to a bacterial reporter gene such as β-glucuronidase (GUS). This construct can be integrated into the plant's genome using a T-DNA vector. Expression of the fused gene can be shown by adding a chromogenic substrate, X-Gluc. Cells that have expressed the fused gene turn blue. If the developmental gene is regulated at the level of transcription, the importance of different parts of the *cis*-active region can be investigated in transgenic plants after fusion with the reporter gene.

4. If the transcription of a developmental gene is regulated, an ectopic expression can give clues to the gene's biological effect. To express the gene ectopically, cDNA of the gene is fused with a strong constitutive promoter, for example the cauliflower mosaic virus (CaMV) 35S promoter. This construct is then integrated into the plant's genome via T-DNA transfer.

---

for altered expression of the promoter–reporter fusion. These kinds of gene regulation studies are suitable for testing and verifying hierarchies of gene effects (that can be explained at the molecular level) which have been deduced by genetic experiments.

3.4 Prospects

Different developmental processes have been investigated using the methods described in this chapter, especially in *Arabidopsis* but also in *Antirrhinum* and *Zea mays* (*see* Chapter 6). Genetic analyses of flower development in both *Antirrhinum* and *Arabidopsis* have been especially successful. These molecular investigations were made easier by the fact that the phenotypically identified genes coded mainly for transcription factors and were expressed only locally (*see* section 6.3.2). Two other developmental processes in *Arabidopsis*, the development of trichomes and pattern formation in the embryo have been analysed, although not yet to the same level of detail as flower development. However, the foundations for molecular investigations of these processes have been laid by genetic analysis (*see* sections 6.1 and 6.2.7).

3.5 Bibliography

Bowman, J. L., Smyth, D.R., and Meyerowitz, E. M. (1991). Genetic interactions among floral homeotic genes of *Arabidopsis*. *Development*, **112**, 1–20.
Genetic analysis has provided an example of how phenotypic analysis of double mutants can lead to a model of the interactions of developmental genes.

Giraudat, J., Hauge, B. M., Valon, C., Smalle, J., Parcy, F., and Goodman, H. M. (1992). Isolation of the *Arabidopsis ABI3* gene by positional cloning. *The Plant Cell*, **4**, 1251–61.
This was one of the first examples of map-based cloning.

Hülskamp, M., Miséra, S., and Jürgens, G. (1994). Genetic dissection of trichome cell development in *Arabidopsis*. *Cell*, **76**, 555–66.
Investigation of genetic mosaics to see if genes for trichome development act autonomously.

Jack, T., Fox, G. L., and Meyerowitz, E. M. (1994). *Arabidopsis* homeotic gene *APETALA3* ectopic expression: transcriptional and posttranscriptional regulation determine floral organ identity. *Cell*, **76**, 703–16.
Different techniques (ectopic expression, reporter gene fusion, indirect immunostaining, *in situ* hybridization) were used in combination to investigate expression and autoregulation of an important gene for flower development.

Jackson, D., Veit, B., and Hake, S. (1994). Expression of maize *KNOTTED1* related homeobox genes in the shoot apical meristem predicts patterns of morphogenesis in the vegetative shoot. *Development*, **120**, 405–13.
Good example of *in situ* hybridization and indirect immunostaining for examining gene expression.

Jürgens, G., Mayer, U., Torres Ruiz, R. A., Berleth, T., and Miséra, S. (1991). Genetic analysis of pattern formation in the *Arabidopsis* embryo. *Development*, Suppl., **1**, 27–8.
Pattern formation in the *Arabidopsis* embryo is used to outline the genetic approach for the analysis of the developmental processes.

Long, D., Martin, M., Sundberg, E., Swinburne, J., Puangsomlee, P., and Coupland, G. (1993). The maize transposable element system *Ac/Ds* as a mutagen of *Arabidopsis*: identification of an *albino* mutation induced by *Ds* insertion. *Proceedings of the National Academy of Sciences of the USA*, **90**, 10370–4.
Insertion mutagenesis used for gene tagging.

Meyerowitz, E. M. (1989). *Arabidopsis*—a useful weed. *Cell*, **56**, 263–9.
Easy-to-read short review about advantages of *Arabidopsis* as a

model organism for developmental studies.

Yanofsky, M. F., Ma, H., Bowman, J. L., Drews, G. N., Feldmann, K. A., and Meyerowitz, E. M. (1990). The protein encoded by the *Arabisopsis* homeotic gene *agamous* resembles transcription factors. *Nature*, **346**, 35–9.

Leading example of isolation of a developmental gene using T-DNA mutagenesis for gene tagging. Proof of the tight linkage by segregation analysis and complementation of mutant alleles, using T-DNA for transformation.

4 The genetic material of plant cells is compartmentalized: structure and expression of the subgenomes

Contents

4.1 The nuclear genome

4.1.1 The nuclear genomes of higher plants differ in size and complexity

4.1.2 Molecular anatomy of the nuclear genome; for example grasses (Gramineae)

4.1.3 Genome organization and families of genes: an overview

4.1.4 The 35S promoter of cauliflower mosaic virus as a typical example of a transcriptional control region of plant nuclear genes

4.1.5 Tissue- and development-specific regulation of anthocyanin synthesis in *Zea mays*—an example of the interaction of several transcription factors

4.1.6 Chromatin organization and DNA methylation influence gene expression

4.1.7 Post-transcriptional control mechanisms

4.2 The plastome

4.2.1 Plastids have their own genomes which are not inherited in a Mendelian fashion

4.2.2 The plastome is polyploid and polyenergid

4.2.3 Plastid DNAs are circular molecules, usually containing two inverse repeats

4.2.4 The repertoire of genes from land plant plastomes

4.2.5 Plastids are semi-autonomous organelles

4.2.6 Plastid genes are typically organized into polycistronic transcription units

4.2.7 Polycistronic transcription units have complex transcription patterns that arise from RNA processing or multiple promoters

4.2.8 Transcription and RNA stability are important determinants of plastid genome regulation

4.3 The chondriome

4.3.1 The chondriomes of most land plants are composed of a dynamic population of DNA molecules

4.3.2 Coding capacity of plant chondriomes

4.3.3 A brief look at mitochondrial gene expression

4.3.4 RNA editing in plant mitochondria and plastids: sequence differences between genes and their transcripts

4.4 **Interactions between different genetic compartments**

4.4.1 Interaction between genome and plastome, using *Oenothera* as an example

4.4.2 Nuclear genes control the expression of plastid genes, but so can plastids affect nuclear gene expression

4.5 **Bibliography**

Preface

Development can be viewed as the result of an ordered interaction of genes that are differentially expressed in space and time. This raises a question—what regulatory mechanisms do plants have to control the activity of their genes? Figure **4.1** shows that the expression of protein-coding genes is a multi-step process. It involves transcription, mRNA production and translation, then finally production of a mature, functional protein. Regulation of gene expression means that at each of these steps it can be determined whether the mature protein is to be produced, and if so, how much should be produced. It could be that the plant regulates gene expression at every step of protein production. However, it turns out that for regulation some steps are more important than others. To control gene activity, gene expression is most frequently regulated at the level of transcription. On the other hand, examples from the animal kingdom have shown that post-transcriptional control (for example, alternative splicing) can be just as efficient and safe a regulatory mechanism.

Genetically speaking, plants are more complex than animals. Their genetic material is divided between three compartments: the nucleus/cytosol, plastids, and mitochondria. Plants therefore have to control not only the many genes in each of these compartments, but must also co-ordinate the expression of genes between the three genetic compartments (Figure **4.2**).

Control of developmental processes is determined mainly by the nuclear genome. We will first deal with the

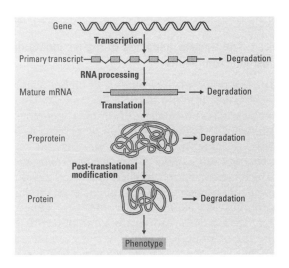

Figure 4.1 The stream of genetic information and its regulation from the gene to the phenotype.

questions: What is the structure of the nuclear genome? What factors does the plant use to control the regulation of its genes? Then, after examining the plastid and mitochondrial genomes, we will look at how the three genomes interact.

Figure 4.2 The genetic material of a plant cell is compartmentalized.

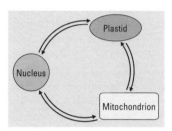

4.1 The nuclear genome

Definition The term *genome* was originally used for genes that are inherited by Mendelian rules, i.e. genes from the nucleus. The *plasmone* was the term used to describe the sum total of all the genetic information from organelles in the cytosol, i.e. the plastids and mitochondria, which do not inherit in a Mendelian fashion. The terms *plastome* and *chondriome* were invented to describe the genes of the plastids and mitochondria, respectively. Nowadays things have been simplified and the term genome is used to describe all collections of genetic information in a cell, no matter what their origin. To avoid any potential misunderstandings, it makes sense for us to refer to genetic information held in the nucleus as the *nuclear genome*. In any case, this is often what is meant when the term genome is used.

4.1.1 The nuclear genomes of higher plants differ in size and complexity

The size of the nuclear genome of higher plants can differ by as much as 200-fold (Table **4.1**). The crucifer *Arabidopsis thaliana* has the smallest of all plant genomes, between 100 and 145 Mb (depending on measurement technique). This is roughly the same size as the genome of *Caenorhabditis elegans* or *Drosophila melanogaster*, both of which are intensively studied model systems for development in the animal world. Lilies have the largest plant genomes—at about 30 000 Mb (the genome size of *Fritillaria assyrica* can be up to 48 000 Mb) they reach the size of amphibian genomes (for example, salamander: 50 000 Mb), the largest of the animal genomes.

Species	Genome Site (Mbp/1C)
Arabidopsis thaliana	145
Oryza sativa ssp. Indica	419–463
Sorghum bicolor	748–772
Lycopersicon esculentum	907–1000
Solanum tuberosum (2n = 4X)	1597–1862
Nicotiana plumbaginifolia	2287
Zea mays	2292–2716
Pisum sativum	3947–4397
Nicotiana tabacum (2n = 4X)	4221–4646
Hordeum vulgare	4873
Triticum aestivum (2n = 6X)	15 966
Tulipa sp.	24 704–30 687
Homo sapiens	3500
Drosophila melanogaster	165
Caenorhabditis elegans	100
Saccharomyces cerevisiae	4
Escherichia coli	4

Table 4.1 The size of the nuclear genomes of angiosperms. The DNA content of intact nuclei has been estimated after staining with propidium iodide in flow cytometry using fluorescence spectroscopy. The DNA content (in picograms) has been converted to the number of base pairs. 1 picogram (pg) equals 985 million base pairs (Mbp) The symbol 1C indicates the DNA content of a haploid genome equivalent in the unreplicated state. (Data for the plant genomes have been taken from the list of Arumuganathan and Earl, *Plant Mol. Biol. Rep.*, **9**, (1991), 208–18.)

Why do plant genomes vary so much in size? DNA reassociation experiments have provided much information about this. Such experiments enable the frequency of occurrence of a single sequence in a genome to be estimated, and thus determination of the complexity of genome components. In general, three separate genome components can be differentiated: highly repetitive, moderately repetitive, and single copy sequences. The following examples of *Arabidopsis thaliana* and *Zea mays* demonstrate this kind of genome analysis (Figure **4.3**).

There is an approximately twentyfold difference in size of the nuclear genomes of the two plants (Table **4.1**). Whereas the genome of *Arabidopsis* is composed of 80 per cent single-occurrence and 20 per cent medium- or highly repetitive sequences, *Zea mays* has mainly repetitive sequences; only one-third of the maize genome is composed of single-occurrence sequences. For this reason, *Arabidopsis* is a better candidate for genome mapping or complete genome sequencing.

Why it should be necessary for a nuclear genome to have a large number of repetitive sequences is unclear. Most of the repetitive sequences seem to be unnecessary for flowering plants. The evidence for this is that both *Arabidopsis* and *Z. mays* have the entire repertoire of metabolic and differentiation functions that are typical of higher plants.

Differences in genome size cannot be explained just by different amounts of repetitive sequences. The complexity of the single-occurrence genes also differs greatly (Figure **4.3**). Polyploidy could be responsible for this. Autopolyploidy is rare; often alloploidy is the norm, as for

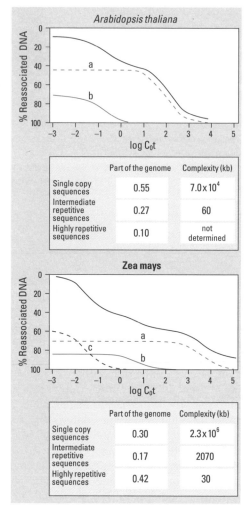

Figure 4.3 Reassociation kinetics of *Arabidopsis thaliana* and *Zea mays* genomic DNA. All the cellular DNA was sheared to give an average size of 300–500 bp, then heated so that it became single stranded as the two strands disassociated. After this the speed of reassociation of complementary DNA sequences was measured. Single- and double-stranded DNAs were separated using chromatography on a hydroxyapatite medium. Using computer analysis the experimentally deduced reassociation kinetic (solid black line) was divided into three kinetic components. The quickly reassociating component with a low C_0t value (c; in the case of *A. thaliana* this was not estimated kinetically) represents the highly repetitive sequences. Intermediate repetitive sequences are represented by curve b and single copy sequences by curve a. Using these reassociation kinetics we can estimate the share of

these components in the genome, as well as their complexity. Since total DNA has been used, the DNAs from the plastids and mitochondria are included. In plants the mitochondrial DNA accounts for about 1–2 per cent of the total genome, so this DNA can be discounted. However, depending on plant species and on the particular plant tissue, the plastid DNA can represent up to 30 per cent of the total DNA of a cell. Therefore plastid DNA can be found in the intermediate repetitive component. The majority (*c.* 2/3) of the intermediate repetitive sequences of *Arabidopsis thaliana* consist of plastid DNA. Therefore the share of single sequences in the nuclear genome is about 80 per cent. (According to Leutwiler, L. S., Hough-Evans, B. R., and Meyerowitz, E. M. (1984). *Molecular and General Genetics,* **194**, 15–23; and Hake, S. and Walbot, V. (1980). *Chromosoma,* **79**, 251–70).

instance in *Nicotiana tabaccum*. This plant is allotetraploid, and was formed from *Nicotiana sylvestris* and *Nicotiana tomentosiformis*. The genome of the hexaploid seed wheat *Triticum aestivum* was formed when three different genomes combined.

Differences in the complexity of single-copy sequences occur not only through polyploidy but also by amplification or deletion of single-copy genome regions. An example of local amplification is the *rbcS* family that codes for the smaller subunit of ribulose-1,5-bisphosphate carboxylase/oxygenase (Rubisco). In tomato (*Lycopersicon esculentum*) there are five *rbcS* genes at three different loci. Two of the loci have only one gene each, but the third has three genes ordered in tandem, and these three genes are more similar to each other than to those at the other two loci. The genes at the third locus are therefore the result of gene amplification.

Finally, differently sized introns can add to the size differences of single-copy sequences. For example, the gene for enolpyruvyl-shikimate-3-phosphatase, a key enzyme in the shikimate pathway, is three times as big in petunias as it is in *Arabidopsis* because of the sizes of its introns.

Even though we now know several reasons for the differences in complexity of single-copy sequences in plants, it is yet to be discovered how genome differences like this affect metabolism and differentiation, or the evolutionary potential of a plant.

4.1.2 Molecular anatomy of the nuclear genome; for example grasses (Gramineae)

DNA reassociation experiments give only a limited insight into a genome's structure. Their biggest failing is that they impart no information about the anatomy of the genome, i.e. which genes are present and how they are arranged. The first insights into this anatomy were given by classical genetic techniques. Genes whose phenotypes can be defined because of mutations can be sorted into linkage groups by crossing experiments. The ways the genes are arranged in the linkage groups can then be estimated using the frequency of recombination. Using these techniques it was possible to establish genome maps that were very nearly exact for several plant species (*Zea mays*, *Lycopersicon esculentum*, and *Arabidopsis thaliana*). However, this method is rather time-consuming. Other limitations are that only genes with distinct phenotypes can be mapped, and the resolution of the maps is limited because there are never as many phenotypical markers available as one would wish. Only the advent of molecular markers such as restriction fragment-length polymorphism (RFLP; *see* section 3.2.2 for more details) have made it possible to produce high-resolution maps.

Genetically, molecular markers behave like 'normal' loci. They provide a framework of orientation points in the genome. Genes that have been cloned as cDNAs or genomic clones can be mapped directly. If they are single-copy genes and they can also be found in the genomes of other plant species by Southern hybridization, they provide an ideal tool with which to compare the structures of different plant genomes.

This sort of genome comparison has been applied to rice (*Oryza sativa*) and maize (*Zea mays*). Both are grasses (family Gramineae), but are in different subfamilies: rice belongs to the Bambusoideae, maize belongs to the Panicoideae. According to phylogenetic analyses, these species formed from a common ancestor about 50 million years ago. As shown in Table **4.1**, the maize genome is six times the size of the rice genome. Despite the differences in genome size and the long evolutionary separation, RFLP mapping with cDNA probes shows that substantial parts of the genome are still colinear (Figure **4.4**). This means they have the same linear sequence with respect to the marker loci that were used for mapping. This phenomenon is also known as genome synteny. Throughout the time that the two species have been evolving independently of one another, there have been no major rearrangements to their genomes. Another surprising result stems from the genome comparison: the maize genome is tetraploid, not diploid. Probes that only recognize a single locus in rice will generally recognize two loci in maize. For example, the whole of rice chromosome number 4 occurs in chromosomes 2 and 10 of maize (Figure **4.4**).

In the meantime, it has been shown that colinearity of large stretches of genomes happens not only between maize and rice but also between these species and wheat (*Triticum aestivum*), a member of the subfamily Pooideae. The results from mapping one grass species can therefore be applied directly to another; the grass family is, in effect, a single enormous genetic system. Detailed knowledge of the genome organization of related species is of importance in helping us to understand the molecular basis of development from an evolutionary point of view.

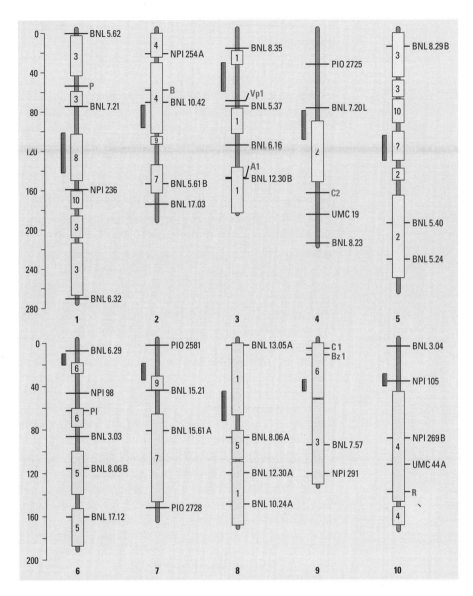

Figure 4.4 Genome colinearity in *Zea mays* and *Oryza sativa*. Using a population of recombinant inbred lines, an RFLP map has been established for maize. RFLP probes from maize varieties from different origins (BNL, NPI, PIO, and UMC) and single cDNA probes from rice (not shown) have been used for hybridization. On the linkage groups of maize (chromosomes 1–10) the parts of the rice chromosomes (rectangles with chromosome numbers) that have the same arrangements of marker loci are indicated. The dark red-shaded rectangles to the left of the maize chromosomes mark the approximate area of the centromere regions. The symbols in red text (P, Pl, B, Vp1, A1, C2, C1, Bz1, and R) mark positions of structural and regulatory genes for anthocyanin synthesis in maize (*see* section 4.1.5). The scales at the left of the picture show the genetic distances measured in centiMorgans. (According to Ahn, S. and Tanksley, S. D. (1993). Proceedings of the National Academy of Sciences USA, **90**, 7980–4.)

4.1.3 Genome organization and families of genes: an overview

The nuclear genes of plants have the exon/intron structure typical of multicellular eukaryotes. The number of exons and introns varies from gene to gene, but is usually identical for the same gene in different species. The sequence flanking a gene at the 5′ end usually controls transcription of the gene and is called a *promoter*. These can be very variable. Some sequence motifs are often found to be conserved in different species, and probably have a functional role in regulating the gene at the level of transcription. For example, many genes have a so-called *TATA box* close to the point where transcription begins, which is necessary for establishing the transcription initiation complex (*see* Figures **4.6** and **4.8**).

Definition The term *promoter* describes sequences that are either close to or next to the point where transcription starts and which have a fixed location and arrangement relative to the transcription initiation site. Elements that do not have fixed positions and polarities in relation to the transcription initiation site and which can act from a distance are called *enhancers*. Enhancers can also be located within the transcribed area or behind the coding region. Enhancers can be simple constructions or can have multiple binding sites for *transcription factors*. Sometimes, distinguishing between promoters and enhancers can be difficult, so often 'promoter' is used to describe the complete region that controls transcription.

At the 3′ ends of genes of lower and higher animals, the polyadenylation signal sequence AAUAAA can often be found. Only 30 per cent of plant genes have this signal. This leads to the conclusion that the *cis*-acting signal needs different features in plants than in animals. Also, *in vivo* experiments in yeasts using the 3′ regions of plants have hinted at functional differences between the polyadenylation signals of plants and animals.

Many plant genes are members of *gene families*. We have already mentioned the *rbcS* gene family, which has three members in *Arabidopsis thaliana* and more than 12 members in hexaploid wheat (*Triticum aestivum*). RbcS proteins taken from the same species are nearly identical, with no more than four amino acid differences between them. They differ only in the strengths and specificities of their expression. Whether minor differences in primary structure lead to functionally different proteins is unknown, but seems unlikely.

On the other hand, the *CAB* family has a clear diversity. *CAB* genes code for the chlorophyll-*a/b*-binding proteins that go to make the peripheral antennae in photosystems I and II that harvest photons of light. Each *CAB* protein clearly differs in primary structure from the others, but each also has a different function in the light-harvesting complex.

Members of a single gene family can also have functions in completely different areas. The phosphoenolpyruvate carboxylases are good examples of this. In C_4 and crassulacean acid metabolism (CAM) plants they catalyse primary fixation of CO_2 during photosynthesis. They also participate in the basic metabolism of all plants. Thus the different phosphoenolpyruvate carboxylase genes not only differ in their expression patterns, but also the enzymes themselves differ in their kinetics and regulation.

The importance of gene families for plant survival is clear. If there are several genes with different expression patterns for proteins which have similar or identical functions, the plants can react better to changing circumstances, for instance a change in their environment. Through gene families they gain a phenotypical plasticity that confers on them an evolutionary advantage.

4.1.4 The 35S promoter of cauliflower mosaic virus as a typical example of a transcriptional control region of plant nuclear genes

To express nuclear genes differentially at the right time and place, transcriptional regulation is the most important type of control. Controlling gene expression at the transcriptional level necessitates that regulatory signals in the DNA interact with *trans*-active proteins and thus alter the frequency of transcription initiation. To understand the basic mechanisms, the recognition sites in the control regions of the genes and the associated *trans*-acting factors need to be identified (Box 4.1).

A description of the basic structure of a plant promoter follows, using cauliflower mosaic virus 35S as an example. This promoter was named after its main transcript, the viral 35S RNA, and is the most extensively studied plant-active promoter to date. Detailed analysis of this promoter has led to basic, but generally valid, knowledge of the structures and functions of plant promoters. For this reason, we will concentrate on this promoter instead of other, less well-described, plant promoters. The 35S promoter is used for both basic fundamental research and biotechnology because it is expressed strongly and constitutively.

Box 4.1 How is the transcription control region of a gene characterized? (*see* Figure 4.5)

1 Likely transcription starting points are examined.

Using primer extension or S1 nuclease protection experiments followed by sequencing, the 5′ end of the RNA can be determined. By convention, the 5′ end of the RNA is believed to be the starting point of transcription. However, this is not necessarily always the case because the 5′ end of an RNA can be generated by RNA processing.

2 Deletion analyses of the transcription control region can identify important cis elements.

First, the presence of a complete transcription control region within the available cloned sequences has to be determined. To do this, the 5′ regions are fused to a reporter gene, for example the *β*-glucuronidase gene (GUS) from *E. coli*, and the product is transformed stably into test plants. But *cis* control regions can also lie within the transcribed region of the 5′ untranslated region, so this region should also be included in the test sequence. If the transcription profile of the reporter gene is the same as that of the original, one can be sure to have isolated the entire transcription control region. Otherwise, a search for further *cis*-acting regions in the introns or the 3′ regions flanking the gene will be necessary.

Sequence deletions are normally made starting from the 5′ end (Figure 4.5). If necessary, deletions will also have to be made from the 3′ end of the region, although in this case a shortened, heterologous TATA region must be fused to the 3′ region flanking the shortened construct, so that the sequences necessary for the assembly of the transcription initiation complex are available. A linker scanning mutagenesis can be used to complete analysis of the transcription control region. With this kind of mutagenesis, the complete region is investigated using 10–20 bp deletions to scan for *cis* elements. After this, directed mutagenesis can reveal the function of a putative *cis*-regulatory sequence motif.

The expression characteristics of deletion mutants produced, as outlined above, should be investigated in intact plants. Only in this way can one be assured that the apparent development and tissue specificity of the expression process is correct. Unfortunately, pressures of work, and the desire to save time, often lead to transient expression systems being used as convenient, less labour-intensive, shortcut methods. Strictly though, it is still necessary to have at least one test case where the researcher can make sure that results gained from transient expression systems accurately reflect those from *in vivo* systems.

3 Gel retardation and footprint experiments can help identify binding sites of trans-regulatory factors.

The sequence motifs, previously shown in deletion and mutagenesis experiments to be essential for normal transcription, can be analysed *in vitro* by gel retardation and footprint experiments for possible binding sites for *trans*-regulatory factors (*see* Figure 4.5). Then *in vivo* footprinting can be used to test, in the genomic context, whether the binding sites that were identified *in vitro* are also utilized in intact cells.

4 Using the binding site as a probe, trans-regulatory factors can be isolated from cDNA expression libraries.

The binding site can be used as a probe to isolate matching clones for *trans*-regulatory factors from a cDNA library. This kind of experimental approach is based on the observation that the DNA binding domain of a transcription factor often acts independently of the other domains in the protein, and so can act as part of a fusion protein. To implement this, a cDNA library is expressed in a *β*-galactosidase expression vector (lambda gt11, for example), so that the cDNA sequence of interest is expressed as a fusion protein. This method will fail if the transcription factor being examined only recognizes its own binding site when it is complexed with other proteins.

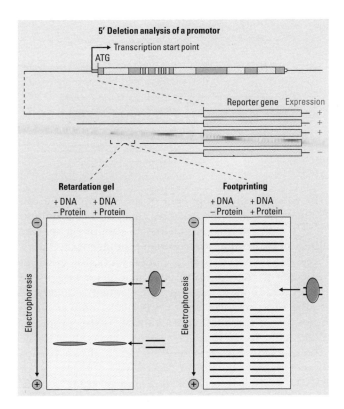

Figure 4.5 Analysis of a transcription control region (*see* Box **4.1** for explanation).

Deletion experiments (*see* Box **4.1**) have shown that the 35S promoter possesses two main domains. Domain A is mostly responsible for the promoter's activity in roots, especially the peripheral roots and the pericycle region. It also regulates activity in the apical meristem of the shoot.

The expression pattern of domain B is more complex. It causes strong expression in plant parts that are above the ground, such as the leaves and the vascular region, but is also slightly active in other tissues. This broad expression profile indicates that domain B is composed of several sub-domains. They have their own characteristic expression specificities and interact with each other, but also interact with domain A.

The 35S promoter is thus built up from single modules, and the effect of combining them is not merely additive, but synergistic. Such modular construction is typical for many nuclear plant promoters, but the majority of these have much more limited expression patterns than the viral 35S promoter. In general, their activity is limited to certain cells or tissues and can be controlled specifically by endogenous or exogenous signals.

Figure 4.6 Structure of the 35S promoter of the cauliflower mosaic virus.The domains A and B and their subdomains are marked by different colours. The TATA subdomain contains the TATA element. The numbers of the nucleotides are related to the arrow which marks the beginning of transcription. The sequence from -208 to -46 is called the enhancer because it retains its function when it is placed with reversed polarity in front of the TATA region. This illustrates well how difficult a distinction between the terms promotor and enhancer can be. The location and the sequence of the *cis*-regulatory elements *as-1* and *as-2* which interact with the transregulatory activities ASF-1 or ASF-2 are shown. The arrows on top of the sequences mark the typical motifs of these binding sites. The ASF-1 and ASF-2 binding activities were demonstrated with gel retardation and footprinting techniques using crude nuclear extracts of *Nicotiana tabacum*. (From Benfey and Chua 1990).

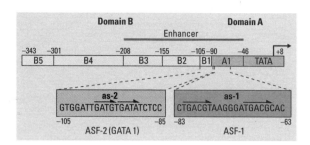

The root-specific component of the 35S promoter can help to explain how the *cis*-active elements of a promoter interact with *trans*-active factors to produce a specific pattern of expression. Deletion and mutagenesis experiments have shown that the *as-1* sequence element (Figure 4.6) is required and is sufficient for activity of domain A in roots. If this sequence element is integrated into the *rbcS* promoter, which is normally only active in photosynthetic tissues, the resulting hybrid promoter will be active in the root.

The *as-1* sequence element is composed of two TGACG modules arranged in tandem, and the ASF-1 nuclear factor binds to it. This factor can be found in crude protein extracts from isolated nuclei, and can be shown by retardation and footprinting experiments (*see* Box 4.1 for methods). Using the binding site as a probe, a cDNA has been isolated, the protein product of which, TGA1a, has the same binding specificity as the ASF-1 nuclear factor. TGA1a and ASF-1 are obviously identical, or at the least TGA1a seems to be most important component of the *trans*-regulatory activity of ASF-1. TGA1a belongs to the basic leucine zipper class of proteins (Table 4.2) that are fairly widespread in animals and plants, and form homo- or heterodimers.

TGA1a is expressed mainly in the roots. In the leaf, its expression is 10 times less efficient. Why then is domain A of the 35S promoter not active in the leaf as well? Is root-specificity perhaps a quantitative phenomenon? That is, only in the root is the concentration of TGA1a sufficiently high to compete with other promoters and supply the *as-1* binding site with TGA1a. Two kinds of experiments have confirmed this hypothesis. If domain A has four extra *as-1* elements, it is active in the leaf. The same result occurs if the concentration of TGA1a in the leaf is increased. This demonstrates that promoter specificity can be the result of the quantitative interactions of *cis* elements and *trans*-regulatory factors.

Classification	Examples	Function	Species
bZIP proteins	TGA1a	*as-1*-dependent expression	*Nicotiana tabacum*
	OPAQUE 2 (o2)	Biosynthesis of the 22 kDa zeins	*Zea mays*
bHLH proteins (Myc-like)	R	Anthocyanin synthesis	*Zea mays*
	B	Anthocyanin synthesis	*Zea mays*
Myb proteins	C1	Anthocyanin synthesis	*Zea mays*
	Pl	Anthocyanin synthesis	*Zea mays*
	GL1	Trichome development	*Arabidopsis thaliana*
Proteins with homeodomains	KNOTTED 1 (Kn1)	Meristem function	*Zea mays*
	BELL 1	Ovule formation	*Arabidopsis thaliana*
	GLABRA 2 (Gl2)	Trichome development	*Arabidopsis thaliana*
	Zmhox 1a	Metabolic regulation of the *shrunken* gene	*Zea mays*
Zinc finger proteins	CONSTANS (CON)	Flower induction	*Arabidopsis thaliana*
	Prolamine-box binding factor (PBF)	Zein synthesis	*Zea mays*
AP2 proteins	APETALA 2 (AP2)	Flower formation	*Arabidopsis thaliana*
	AINTEGUMENTA (AIN)	Ovule formation	*Arabidopsis thaliana*
MADS box proteins	DEfiCIENS (DEF)	Flower formation	*Antirrhinum majus*
	GLOBOSA (GLO)	Flower formation	*Antirrhinum majus*
	AGAMOUS (AG)	Flower formation	*Arabidopsis thaliana*
Not yet classified	GT-1	Light regulation	*Nicotiana tabacum*
	GT-2	Light regulation	*Oryza sativa*
	Viviparous 1 (Vp1)	Seed ripening	*Zea mays*
	COP1	Light regulation	*Arabidopsis thaliana*

Table 4.2 A selection of plant transcription factors. The leucine zipper (as a dimerization domain) and a basic DNA binding motif are the characteristics of bZIP proteins. Instead of the leucine zipper the bHLH proteins contain a helix–loop–helix motif as a dimerization domain. These and an additional leucine zipper are characteristic of the Myc proteins (*myc* is a human proto-oncogene). The class of Myb proteins is also named after a human proto-oncogene. The homeodomain is a DNA-binding motif that was first found in homeotic proteins of *Drosophila*, and that in its structure is similar to the helix–turn–helix motif. The name 'MADS box proteins' is derived from the combination of the initials of the following transcription factors: MCM1 (mating-type-specific transcription of yeast), AGAMOUS, DEFICIENS (both homeotic proteins of flower morphogenesis), and the Serum response factor (which in mammals regulates, together with p62, the transcription of C-*fos*). The zinc finger proteins contain DNA-binding domains with cysteine and histidine residues which complex Zn^{2+}. The shapes of the different zinc finger proteins are not uniform. The transcription factor Sp1 and the glucocorticoid receptor are examples of different types of zinc finger proteins of mammals. The APETALA2 protein lends its name to a new group of plant transcription factors which are characterized by a 17 amino acid motif. According to structural predictions, this motif contains an α-helix (the AP2 box).

4.1.5 Tissue- and development-specific regulation of anthocyanin synthesis in *Zea mays*—an example of the interaction of several transcription factors

The 35S promoter is an example of how organ-specific gene expression results from the equilibrium between the concentrations of transcription factors and the related *cis*-acting regulation signal. The following example uses anthocyanin synthesis in *Zea mays* to show how, with the interactions of several *trans*-regulators, the genes of a metabolic pathway can be regulated in tissue- and development-specific ways. The unravelling of the regulatory mechanisms in this system also turns out to be a good example of how, just by combining genetic and molecular experiments, the interactions of the single *cis*- and *trans*-acting factors can be understood at the mechanistic level.

Biosynthesis of anthocyanins and phlobaphenes starts with a common origin, malonyl- and coumaroyl-CoA, and later separates at the flavanone level. Chalcone synthase (C2) and an NADP-dependent flavanone/dihydroflavonol reductase (A1) are enzymes common to both synthesis pathways. However, for example, UDP-glucose flavonol-(*O*)-3-glucosyltransferase (Bz1) is unique to anthocyanin synthesis (Figure **4.7**). Phlobaphenes are found mainly in the mature pericarp and spike. Anthocyanins are synthesized in many different organs of *Zea mays*, but there are many variations in different maize varieties. Six genes, *C1*, *Pl*, *R*, *B*, *P*, and *Vp1* are important in its regulation. All code for transcription factors (Figure **4.7**). The R, B, C1, and Pl proteins all have similar functions. C1 and Pl belong to the *myb* gene family (*see* Table **4.2**). C1 is required for anthocyanin synthesis in the seed (i.e. in the aleurone layer and embryo). The transcription factor Pl, which is homologous to C1, has the same function throughout the plant. The expression of both transcrip-

Figure 4.7 Genetic control of anthocyanin biosynthesis in *Zea mays*. For explanation see text. (According to Doebley, 1993).

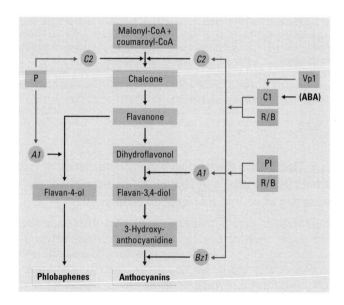

tion factors is tissue-specific. Activity of *C1* or *Pl* alone is not sufficient to start anthocyanin synthesis—R and B must also be present. R and B are homologous proteins, both possessing a basic helix–turn–helix domain as found in the sequence of the proto-oncogene *myc* (*see* Table **4.2**). Whereas B is coded for by a single gene, R proteins are encoded by a family of genes. The *R* and *B* genes are not expressed in a tissue-specific fashion, as are *Pl* and *C1*. Depending upon maize variety, they show different expression patterns (in terms of time and location of expression). This is especially the case with members of the *R* gene family, and explains the broad variety of pigmentation patterns in maize varieties.

C1 and B/R are necessary for transcribing the structural genes of anthocyanin synthesis. This has been proved by both genetic investigations and transient co-expression of both groups of transcription factors in cell cultures and intact maize seeds. The *Bz1* and *A1* promoters will only be active if equivalent amounts of C1 and R/B are present. Both promoters apparently bind as homodimers to their target sequences and in combination are synergistic.

Whereas anthocyanin synthesis requires the activity of both C1/Pl and R/B, phlobaphene synthesis needs only the transcription factor P. P, like C1, is a member of the *myb* family. On its own, P can activate expression of *A1*, but not *Bz1*. In this way, the synthesis of anthocyanins and phlobaphenes is regulated separately.

Anthocyanin synthesis in the seed is regulated not only by *C1* and *R/B*, but also at a higher level by *Vp1*. *Vp1* (*viviparous 1*) is a transcription factor that has no homology to any other transcription factors characterized to date. Together with other genes, it controls the maturing of the embryo (*see* section 5.6.2). As well as other pleiotropic effects, a mutation in *Vp1* can switch off anthocyanin synthesis in the seed, but has no effect on anthocyanin synthesis throughout the rest of the plant. The absence of anthocyanins in seeds of *vp1* mutant plants is due to a lack of C1, the gene for which is switched off in *vp1* mutant plants. *C1* transcription is therefore controlled by *Vp1* and in this way anthocyanin synthesis is integrated into a higher-level developmental programme.

The example of anthocyanin synthesis in maize has introduced us to a two-level genetic hierarchy of transcription control. There can, however, be many cascading levels of control, as seen in *Drosophila*. The participating transcription factors are produced in sequence by activation of the corresponding genes. Active transcription factors originate from protein modification, as well as from *de novo* synthesis. Many transcription factors have domains for dimerization, and are present in the cell as homo- or heterodimers. Acquiring, discarding, or exchanging an interacting partner molecule can also alter the activity of a transcription factor. In this way, eukaryotic organisms are able to transcribe genes separately, or to co-ordinate when and where several genes are transcribed.

4.1.6 Chromatin organization and DNA methylation influence gene expression

Regulation signals and their associated *trans*-active proteins operating at the level of DNA are not the only regulators of nuclear gene expression. Nuclear DNA is always packaged as chromatin, and this provides another opportunity for gene regulation. The chromatin is not a uniform thread of nucleosomes; it is separated into discrete parts. There is some evidence that these chromatin domains are in fact large expression structures of one or more genes that are being actively transcribed. In addition, gene activity can also be modified by reversible methylation.

Genes that are being transcribed actively have a different chromatin structure in comparison to inactive genes. This can be demonstrated by digestion experiments using nucleases such as DNAase I or *Micrococcus* nuclease. Active genes possess the same number of nucleosomes as they have when they are inactive, but their chromatin structure seems to be more open and more accessible to the nucleases. Digestion experiments using low concentrations of nuclease show that susceptibility to nuclease digestion is unevenly spread across the gene; some regions are hypersensitive. Nuclease-hypersensitive sites are only 'visible' in the chromatin of cells where the investigated gene is active. They are found especially in the 5′ region of a gene, and mark the DNA sequences with information required for transcription (Figure **4.8**). They can represent binding sites of transcription factors in promoters or enhancers. Hypersensitive sites can also be found near scaffold attachment regions (Figure **4.8**).

Definition *Scaffold attachment regions* (SAR; also known as *matrix attachment regions*, MAR), are defined as DNA sequences that bind specifically to the proteins of the nuclear matrix during an *in vitro* test. SARs are AT-rich and can be several hundred bases long, and it is difficult to find any consensus sequences in their sequence motifs. It is assumed that SARs identified *in vitro* have similar functions

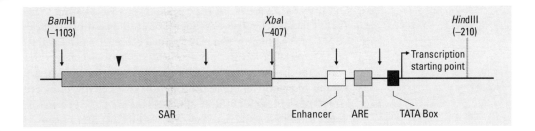

in vivo, i.e. they anchor chromatin to the nuclear scaffold. Whether such areas of chromatin bordered by SAR sequences are also functional expression domains has yet to be determined.

In mammalian genes, hypersensitive sites have been discovered far away from the 5' and 3' ends of the gene. In the human β-globin gene locus they are found 10–20 kb away from the ε-globin gene. They mark the *locus control region*. This region is essential for the expression of all genes in this locus. Whether such gene-spanning control elements also exist in plants is as yet unknown.

Expression studies with transgenic plants have shown that they, too, have expression-specific structures at the chromatin level. The amount of expression of a transgene in plants that have been similarly transformed, but independently of one another, can differ by orders of magnitude. These fluctuations in expression level (and also in expression pattern) are thought to be due to position—the genome context at the site of transgene insertion can influence transgene expression. For instance, a genome area that is 'silent' might depress transgene expression, or if an enhancer is near the insertion point, expression may be increased, or influenced in a qualitative fashion. Bordering elements that 'guard' a gene or group of genes against their neighbouring genes are known from experiments with transgenic *Drosophila* and mice. Such border elements can be part of the SAR, but do not have to be. The first experiments to elucidate the effects of border elements on transgene expression have been done, but more detailed characterization of these elements will be necessary in the future.

Active genes not only differ from their silent cousins by their chromatin structure; they also differ by the amount of

Figure 4.8 Nuclease-hypersensitive sites in the transcription control region of the *Adh1* gene of maize. The *Adh1* gene encodes ethanol dehydrogenase which is expressed in the roots after anaerobic stress. Arrows mark the DNAase-I-hypersensitive regions and arrowheads the S1-nuclease-hypersensitive sites. The position of the TATA box, the anaerobic response element (ARE), the proposed enhancers, and the scaffold attachment region (SAR) are indicated. The function of SAR sequences were established by *in vitro* tests. (According to Avramova, Z. and Bennetzen, J. L. (1993). *Plant Molecular Biology*, **22**, 1135–43).

methylated cytosine in their sequences. Methylated cytosine (5-methylcytosine) is mainly found as a dinucleotide, CpG, and in plants also as a trinucleotide CpNpG. In general, excessive methylation (*hypermethylation*) of certain cytosines in the promoter area correlates with gene activity being blocked, whereas the cytosines of active genes are methylated less, or not at all (*hypomethylation*). For mammals it has been shown that methylated CpG sequences are recognized by specific proteins. If these proteins bind to these groups, the promoter cannot initiate transcription. It is not known whether there are similar repressor proteins in plants.

Clearly, alterations to the degree of methylation is not the cause of, but merely a side-effect of, a change in gene activity. DNA methylation is essential for animal development. This has been shown for mice with a targeted mutation of the DNA methyl transferase gene. However, how methylation affects regulation is unclear. Several of the lower eukaryotes, such as *Saccharomyces cerevisiae*, *Caenorrhabditis elegans*, and *Drosophila melanogaster* survive without methylation, but have otherwise-similar regulation mechanisms at the levels of transcription and chromatin structure.

4.1.7 Post-transcriptional control mechanisms

Post-transcriptional mechanisms provide plants with the next most important level of gene expression regulation. Even though the actual mechanisms are not yet understood, two possibilities for post-transcriptional control are well-documented for plants—mRNA stability and translation.

As in yeasts and mammals, the 3' untranslated regions of the plant's mRNA contain sequence elements that control the stability of the RNA (e.g. AU-rich sequences). If these 3' regions are transferred to other mRNAs, the mRNAs become unstable. It is not yet clear whether shortening the poly(A) tail of plant mRNAs is the event that initiates degradation of the mRNA. In yeasts it can be shown that after the RNA is de-adenylated at the 3' end, the poly(A)-binding protein is no longer able to bind to the shortened poly(A) tail. This leads to degradation of the cap structure at the 5' end of the RNA, allowing the mRNA to be digested by $5' \rightarrow 3'$-acting exonucleases.

Control of translation can be at several different levels. To affect the frequency of translation initiation, the 5' leader sequence is important. For instance, the leader sequence of tobacco mosaic virus (also known as *omega*) has been shown to be an efficient translation control region. In it there are three copies of an eight-base direct sequence repeat and a 25-base poly(CAA) region. The latter determines frequency of translation initiation.

As with GCN4, a yeast transcription factor, open reading frames in the leader sequence of an mRNA can have a negative effect on translation of the following gene. Three such open reading frames are found in the extraordinarily long leader of the Opaque 2 protein, a transcription activator in zein synthesis. Deletion of the leader, or deactivation of its three initiation codons, results in enhanced Opaque 2 activity in transient expression experiments in tobacco protoplasts. Similar results have been obtained in experiments on the 5′ untranslated region of the mRNA of the transcription activator Lc, a member of the R protein family (Table **4.2**).

Control of translation can not only be at the initiation stage but also during elongation of the emerging peptide. *Amaranthus* seedlings that have been grown in the light reduce the synthesis of the small subunit (SSU) of Rubisco by ten- to twentyfold when they are put in the dark, but the mRNA that codes for this subunit only decreases by a factor of two to three. On the other hand, when the etiolated seedlings are returned to the light SSU synthesis increases twenty- to thirtyfold, but the amount of mRNA only increases by two- to threefold. In neither case is there a change in the profile of the polysomes, so the regulation of SSU synthesis must be at the translation elongation level.

Post-transcriptional control mechanisms are important tools that plants use to tailor their gene expression exactly to their needs. They enable plants to react to environmental changes much more quickly and precisely than regulation at the level of transcription alone would permit.

4.2 The plastome

4.2.1 Plastids have their own genomes which are not inherited in a Mendelian fashion

In 1909, shortly after Mendel's rules were rediscovered, Correns described experiments with *Mirabilis jalapa* which could not be explained by Mendelian rules of inheritance. There are different varieties of this plant with either green, white, or green/white striped leaves. In reciprocal crossing of flowers from green and white branches all offspring were green if the flower that was pollinated was on a green-leaved branch. The following offspring were also green. However, if the pollinated flower was on a white branch, the leaves of its offspring would always be white. Thus the colour of the leaves was determined by the mother; this is known as *maternal transmission*.

Baur (1909) also described non-Mendelian inheritance of leaf colour in *Pelargonium zonale*, but in this case reciprocal crossing experiments showed that the offspring could inherit their leaf colouring from either parent. This is known as a *biparental transmission mechanism*. The F_1 plants were green–white chequered and had different branches (green, white, periclinal, or sectored chimera). Baur also observed that the green parts of the leaves contained normal chloroplasts, but the white parts only contained colourless plastids. After seeing this, he hypothesized that the plastids themselves must be carriers of genetic information. He explained the chequered phenotypes in the following way: the genetically different plastids from the father and mother plants are inherited by their daughter cells at random and later on segregate. As far back as 1909, Baur's theory correctly described the transmission of plastids from parent to daughter plants.

Today we know that as well as being maternally and biparentally transmitted, plastids can be paternally transmitted. Plastid transmission in most angiosperms is via the maternal route. Relatively few genera (*Pelargonium*, *Oenethera*, *Hypericum*) transmit their plastids biparentally. Paternal transmission has never been observed in angiosperms, but is common in gymnosperms such as *Pseudotsuga*, *Sequoia*, and *Pinus*.

4.2.2 The plastome is polyploid and polyenergid

Plastids are polyploid and polyenergid organelles. The multiple DNA molecules of plastids are not randomly distributed inside the organelle, but accumulate at discrete regions, for instance at the thylakoid membrane or the inner membrane. These regions are also known as *nucleoids*. Fully differentiated mesophyll chloroplasts of land plants have from 10 to 20 nucleoids, each with between 2 and 20 DNA molecules. The number of copies of plastid DNA, and the number of plastids per cell, are not fixed. They vary between different species, and also vary *within* species depending on the type of tissue and the current developmental stage.

The primary leaf of a barley seedling is a good example to demonstrate this (*see also* Box **4.2** and Figure **4.9**). At the leaf base, the meristem cells have about 15 proplastids, each containing approximately 100 copies of plastid DNA. As the cell differentiates, more plastids are found—up to 60 chloroplasts in a mesophyll cell in the middle of the leaf. At the same time, copies of plastid DNA double to 200. This equates to approximately 12 000 copies per mesophyll cell. In contrast, older mesophyll cells at the leaf tip have around 3000 copies, or 50 per plastid.

4.2.3 Plastid DNAs are circular molecules usually containing two inverse repeats

All naturally occurring plastid DNAs are circular and double stranded. In contrast to mitochondria, only DNA molecules with identical sizes can be found, thus the plastomes are *unicircular*. However, the linear molecules that are found in anther cultures of grasses are an exception. Plastid DNAs from many plant species that have been regenerated *in vitro* often contain large deletions. These can be seen as albino phenotypes.

Plastid DNAs from the majority of land plants (angiosperms, gymnosperms, mosses, and ferns) range in size from 130 to 160 kb. Larger differences can be seen in the plastomes of algae (compare *Codium fragile* and *Chlamydomonas moewusii*, Table **4.3**). However, their sizes do not differ as much as those of mitochondrial genomes.

Two large inverted repeats are typical in the plastomes of land plants (Figure **4.10**). These inverted repeats separate two single-copy regions that are called, according to their sizes, the small and large single-copy regions. The two inverted repeats of land plants are an original feature. During

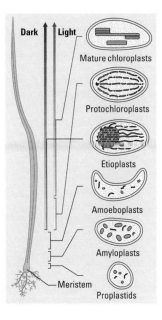

Figure 4.9 Plastid differentiation in leaves of grasses. For explanation *see* Box **4.2**. (According to Wellburn, A. R., Gounaris, I., Owen, J. H., Laybourn-Parry, J. E. M., and Wellburn, F. A. M. (1986). In *Regulation of chloroplast differentiation*, (ed. G. Akoyunoglou, H. Senger), pp. 371–81. Liss, New York.)

Box 4.2 The development gradient in leaves of the Gramineae: a model system for studying plastid differentiation (Figure 4.9)

The leaves of the Gramineae are perfect model systems that can be used for investigating the development of chloroplasts from proplastids. The growth of a leaf of a graminaceous plant begins at the leaf base in the intercalary meristem. Along the leaf axis a gradient of development can be seen. At the leaf base there are proplastids, 2–3 μm in diameter, which, further along the leaf axis, gradually turn into amyloplasts or amoeboplasts.

If the seedling is grown in the light, these plastids will develop into protochloroplasts, and eventually chloroplasts, which are 6–7 μm in diameter. If the plants are etiolated, the proplastids turn into etioplasts. These are characterized by large, semicrystalline prostructures called prolamellar bodies. Etioplasts are devoid of chlorophyll and hence are not photosynthetically active. Only when they are returned to the light do they develop into chloroplasts.

evolution, they have only disappeared from Papilionaceae, two orders of the Geraniaceae (*Erodium* and *Sarcocaulon*), one order of the Scorphulariaceae (*Striga*), and the conifers. Interestingly enough, the inverted repeats of the plastid DNA have also been retained in the parasitic angiosperm species *Epifragus virginiana*. Incidentally, this plant can have extensive deletions and produces pseudogenes (Table **4.3**).

The inverted repeats of land plant plastomes generally range from 20 to 30 kb, but, as in the cases of *Pelargonium* and *Geranium*, they can reach 76 kb (*see* Table **4.3**). Nevertheless, few differences in genome complexity can be found. This is because with the extension of the inverted repeat, no new sequences are produced; only regions that were hitherto unique are introduced into the duplicated segments. Duplications that differ only in size and not numbers of different genes are responsible for the size differences of plastomes of different land plants.

Besides these large inverted repeats, other repetitive sequences are rare. The plastomes of land plants thus have

Table 4.3 The size of selected plastid DNAs and their inverse duplication (according to Palmer 1991).

Taxa	Genome size (in kb)	Inverse duplication (in kb)
Angiospermae		
Nicotiana tabacum	156	25
Spinacia oleracea	150	24
Pelargonium hortorum	217	76
Pisum sativum	120	Not present
Epifragus virginiana	70	22
Oryza sativa	134	21
Gymnospermae		
Pinus	120	Not present
Ginkgo biloba	158	17
Pteridophyta		
Osmunda cinnamomea	144	10
Bryophyta		
Marchantia polymorpha	121	10
Chlorophyta		
Codium fragile	85	Not present
Chlamydomonas reinhardtii	195	22
Chlamydomonas moewusii	292	41
Rhodophyta		
Codium fragile	85	Not present
Chromophyta		
Dictyota dichotoma	123	5

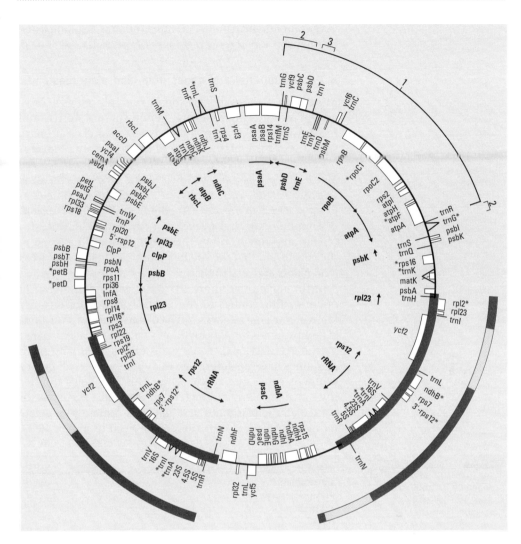

Figure 4.10 Genetic map of the plastid DNA of tobacco. Genes inside the circle are transcribed clockwise. Genes on the outside are transcribed anticlockwise. Arrows inside the circle indicate genes that are transcribed polycistronically. In general operons are named after their first transcribed gene, for example the *psbB* operon. The nomenclature resembles that for bacteria in that the first three letters in lower case name the functional unit, then the components and subunits are named in capitals. *psa*, photosystem I; *psb*, photosystem II; *pet*, photosynthetic electron transport (for example the cytochrome-*b*/*f* complex); *atp*, ATP synthase; *rbcL*, large subunit of Rubisco; *ndh*, NADH dehydrogenase (complex I); *rrn*, ribosomal RNAs; *trn*, tRNAs; *rps*/*rpl*, ribosomal proteins of the small (*rps*) and the large (*rpl*) subunit; *inf*, translation initiation factor; *rpo*, RNA polymerase; *ClpP*, Clp protease subunit; *accD*, acetyl CoA carboxylase; *cemA*, protein of the plastid envelope; *matK*, maturase K. Conserved open reading frames of unknown function are named *ycf* and are numbered. Genes containing introns are marked with an asterix (e.g. *petB*). The thick red bars on the circle represent inverse duplications. The thick segments outside the circle represent inverse duplications in the rice plastome. Parts that are missing in rice are shaded lighter. The numbered brackets (1, 2, 3) outside the tobacco circle mark three overlapping inversion events in which the rice and the tobacco plastome differ. (According to Palmer 1992.)

very compact structures. This observation is also supported by their short intergenic regions of between 10 and 500 bp.

4.2.4 The repertoire of genes from land plant plastomes

More than a thousand plastomes from all representative groups of the angiosperms, gymnosperms, ferns, and mosses have been comparatively mapped. The primary structures of five plastid DNAs of the land plants *Nicotiana tabaccum*, *Oryza sativa*, *Zea mays*, *Pinus thunbergii*, and *Marchantia polymorpha*, and five plastid DNAs of the algae *Chlorella vulgaris* (Chlorophyta), *Euglena gracilis* (Euglenaphyta), *Porphyra purprea* (Rhodophyta), *Odontella sinensis* (Chromophyta), and *Cyanophora paradoxa* (Glaucocystophyta) are known. All of these investigations have led to the conclusion that the coding potentials of plastomes from land plants are very similar.

In the meantime, more than 100 genes and about 30 open reading frames of the plastid DNA of tobacco are known. Most of the genes can be divided into two functional groups: genes for photosynthesis and the gene expression machinery (Table **4.4**). To date, all investigated land plant plastomes have no differences in their repertoires of photosynthesis genes and rRNA genes. This also holds true for ribosomal proteins, although there are some exceptions. In particular, the plastomes of land plants contain a complete set of tRNA genes. Thus there is no need to import tRNA

Table 4.4 The gene repertoire of the tobacco plastome (Kindly provided by B. Stöbe and K. Kowallik, University of Düsseldorf, Germany.)

Function	Number of genes
Transcription	
RNA polymerase	4
Translation	
rRNAs	4
tRNAs	30
Ribosomal proteins	21
Photosynthesis	
CO_2 fixation	1
Photosystem I	5
Photosystem II	14
Cytochrome-*b/f* complex	5
ATP synthase	6
Other	
NAD(P)H dehydrogenase complex	11
Clp protease	1
Acetyl-CoA carboxylase	1
Plastid membrane	1
Maturases	1
Open reading frames	*c.* 30
(ORFs, > 29 amino acids)	

from the cytosol into the plastids (compare this to the mito-chondria, section 4.3.2). These genes have thus been well conserved throughout the evolutionary history of land plants. Some larger differences can be seen in the as yet unidentified open reading frames and the *ndh* genes. For example, *ndh* genes are absent from the plastome of *Pinus*.

4.2.5 Plastids are semi-autonomous organelles

With 120–130 genes and open reading frames, the plas-tomes of land plants have a large number of genes in com-parison to the mitochondria (section 4.3.2). However, from a genetic point of view, plastids are not autonomous. Most plastid proteins are encoded in the nucleus and have to be transported into the plastid after they have been translated in the cytosol. For example, two-thirds of the components that go to make the thylakoid membrane are encoded by the nuclear genome, and only one-third by the plastome.

The dual genetic origin of the plastid is even visible at the level of single protein complexes: the ribosome, Rubisco and the protein complexes of the thylakoid membrane are all genetic mosaics. Successful plastid biogenesis therefore needs the integrated activities of two separate genetic compartments (*see* section 4.4).

4.2.6 Plastid genes are typically organized into polycistronic transcription units

Most plastid genes are arranged as polycistronic transcrip-tion units. Exceptions to this rule are the genes for the large subunit of Rubisco (*rbcL*) and the D1 protein of photosystem II (*psbA*), which are both transcribed monocistronically.

Two features are striking when the organization of plastid and eubacterial genes are compared:

1. In plastids, genes for the subunits of a protein are not organized together into a single operon but are distrib-uted between several transcription units. For example, the 14 plastome-encoded photosystem II genes are dis-tributed between six (monocotyledonous plants) and seven (dicotyledonous plants) different transcription units. In eubacteria, however, genes for the subunits of a protein complex (for example ATP synthase, *unc* operon), or the enzymes of a metabolic pathway (for example tryptophan biosynthesis, *trp* operon) are arranged in one transcription unit.

2. Most polycistronic plastid transcription units are heterogeneous, consisting of genes for several different functions. An example is the *psbB* operon which consists of three genes for subunits of photosystem II and two the cytochrome-*b*/*f* complex (Figure **4.11**). The functional meanings of such combinations are not understood.

4.2.7 Polycistronic transcription units have complex transcription patterns that arise from RNA processing or multiple promoters

Northern blot investigations of polycistronic transcription units usually result in some extremely complex banding patterns. Besides the expected polycistronic transcripts that contain all the genes, there are many smaller RNAs that only make up shared parts of the gene. Such complicated transcription patterns can be attributed to RNA processing, multiple promoters, or multiple terminators.

In the case of the *psbB* operon, it can be demonstrated that RNA processing is the cause of the complex banding patterns.

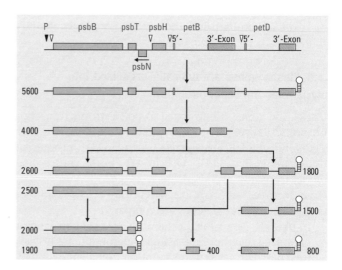

Figure 4.11 The *psbB* operon—RNA processing leads to a variety of oligo- and monocistronic transcripts. *PsbB*, *psbT*, *psbH*, and *psbN* encode the chlorophyll *a* apoprotein CP47, the T-subunit, the 10 kDa phosphoprotein, or the N-subunit of photosystem II, *petB* and *petD* the subunits cytochrome-*b*6 and 4 of the cytochrome-*b*/*f* complex. The letter P marks the site of the only mapped promotor of the *psbB*-operon. The black arrowhead points at the RNA 5′ end that is generated by transcriptional initiation and the white arrowheads at RNA 5′ ends that are due to RNA processing. The complex pattern of transcription has been simplified and shows only completely spliced RNAs. The hairpins at the 3′ end are marked by the usual symbol. The transcript sizes are given in nucleotides. (From Westhoff, P. and Herrmann, R. G. (1988). *European Journal of Biochemistry*, **171**, 551–64.)

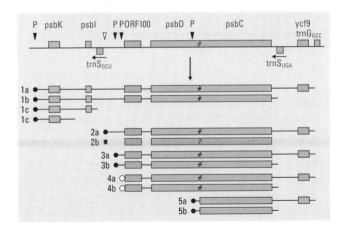

Figure 4.12 The *psbD/C* operon of grasses—existence of multiple promoters. *PsbC*, *psbD*, *psbI*, and *psbK* encode the chlorophyll *a* apoprotein CP43, the D2-protein, or the I and K subunits of photosystem II. The functions of the *ycf9* and the *ORF100* are unknown. It is doubtful whether the *trnG* gene is a part of the *psbD/C*-operon. The letter P indicates promotors. Black arrowheads point at RNA 5′ ends that are due to transcription initiation and white arrowheads point at RNA 5′ ends that are generated by RNA processing. The reading frames of *psbD* and *psbC* overlap by 17 bp. Northern blots of barley revealed that the RNAs 1a–c, 2a–b, 3a–b, and 5a–b are synthesized in etiolated seedlings but that RNAs 4a–b require light and are synthesized by transcription from a light-inducible promotor. (From Sexton, T. B., Christopher, D. A., and Mullet, J. E. (1990). *The EMBO Journal*, **9**, 4485–94.)

A simple scheme showing how these transcripts mature is shown in Figure **4.11**. Endonucleases play an important role in the maturation of the polycistronic transcripts. They make cuts in the intergenic regions of the primary transcript, thus freeing the oligocistronic RNAs. To date, nothing is known about the molecular nature of these endonucleases, but some of their target sequences are known.

In addition to the endonucleases, some exonucleases are also involved in the maturation process, and shorten the RNA by removing portions of it from the 3′ end. Hairpin loops at this end can hinder exonucleases, and so contribute to the stability of the RNA. Clearly though, they are not the sole determinants of the stability of the RNA, as investigations with eubacteria have shown. Such stem–loop structures are associated with proteins in the plastid. The functions of these proteins are essential for the secondary structure as a barrier against exonucleases.

Complex transcription patterns result not only from RNA processing, but also from multiple transcription initiation. A good example of this is the *psbD/C* operon. In grasses this operon contains four photosystem II genes and two open reading frames (Figure **4.12**). Besides the promoter in front of the first gene (*psbK*) there are three further promoters within the operon. One of these promoters is even within the coding region of the *psbD* genes. This operon is also an example of how transcription units with multiple internal promoters might have developed. Two of the operons that are usually separated in land plants, *psbK–psbI* and *psbD–psbC–ycf9*, are combined as a single transcription unit in grasses by virtue of genome rearrangements (*see* Figure **4.10**). The promoters have been retained.

4.2.8 Transcription and RNA stability are important determinants of plastid genome regulation

Expression of plastid genes can be controlled by transcription, RNA stability, and translation. Just how much each of these factors influences regulation depends on the type of plastid, current state of development, and can also vary from gene to gene. Therefore, only a few basic principles will be illustrated in the following paragraphs.

The transcription machinery of the plastid at first seems to be remarkably similar to the equivalent prokaryotic machinery. Many of the plastid gene promoters identified to date resemble the classical *E. coli* promoters that are recognized by the sigma-70 factor. As with prokaryotic promoters, plastid promoters have a conserved region from '−10' to '−35' (Figure **4.13**) which is important for recognition of the promoters and which determines the strength of the promoter.

The so-called PEP promoters (Figure **4.13**) are recognized by a prokaryotic-type RNA polymerase (Plastid-Encoded Polymerase). The genes for the subunit of the core enzyme (*rpoA*, *rpoB*, *rpoC1*, and *rpoC2*) are located in the

PEP promoters:

	'35'			'10'	
rbcL	TACGGTTGGG	TTGCGC	CATATATATGAAAGAGTA	TACAAT	AATGATGT
atpB	AAATTTACTC	TTGACA	GTGGTATATGTTGTATATG	TATATC	CTAGATGT
psbA	TAGATATTGG	TTGAGA	CGGGCATATAAGGCATGT	TATACT	GTTGAATA
psbB	TCAACTCCCA	TTGCGT	ATTGCTACTTATCGAGTA	TAGAAT	AGATTTGT
E. coli	......	TTGACA	17 bp	...... TATAAT	

NEP promoters:

ycf2	ATGTAGATGATGATATCTATACAGATGCATCTTATATATATC	GTAGAAT	GAA
atpB	AATCGTAATAGAAATAGAAAATAAAGTTCAGGTTCGAATTCC	ATAGAAT	AGA
rps2	TCTGCATTTGGTATGGTTATTTGCTTTGGTAATAAAAAGAAT	ATTGAAT	AGA
accD	ATATTATTTTAAATAATATAAAGGGGGGTTCCAACATATTAAT	ATATAGT	GAA

Figure 4.13 Structures of plastid promoters. The areas that resemble the '−10' and '−35' promoter elements of the eubacterial sigma-70 promoters of *E. coli* are underscored in red. These are called PEP promoters because they are driven by the plastid-encoded RNA polymerase. The structures of promoters that are only recognized by the nuclear-encoded RNA polymerase (NEP) are not understood. A possible functional element of these promoters is highlighted in red. Transcription initiation sites are underlined. (According to Hajdukiewicz, P. T. J., Allison, L. A., and Maliga, P. (1997). *The EMBO Journal*, **16**, 4041–8.)

plastome, while the sigma factors are encoded by the nuclear genome. Meanwhile, several sigma factor sequences of red algae and higher plants have been identified. That there are plastid RNA polymerases other than the plastid-encoded RNA polymerases was shown long ago by investigations on the parasitic angiosperm *Epifragus virgiana* which has no plastid-encoded *rpo* genes. The same conclusions were drawn from studies of the barley mutant *albostrians* which has no plastid ribosomes. This evidence has recently been proved by knockout mutagenesis of the *rpoB* gene of tobacco—even though their plastid-encoded RNA polymerase is missing, transcripts of ribosomal RNAs, proteins, and non-photosynthetic proteins such as the Clp protease and the AccD protein can be found in their plastids (*see* Table **4.4** and Figure **4.10**). Only the transcripts of the subunits of photosystems I and II are completely absent from the tobacco mutants. The genes for these transcripts are therefore transcribed exclusively by the plastid-encoded RNA polymerase. The second RNA polymerase of the plastid is called the *nuclear-encoded RNA polymerase*, because of where it is encoded, and its promoters are called *NEP promoters* (Figure **4.13**). The gene for the nuclear-encoded RNA polymerase of the plasmid has recently been isolated from *Arabidopsis thaliana*. The primary translation product contains a transit sequence, which directs the protein into the plastids. The nuclear-encoded RNA polymerase has a very similar amino acid sequence to the mitochondrial RNA polymerase. The RNA polymerases of both organelles therefore appear to have a common evolutionary origin. They are related to the single-chain enzymes of the T3 and T7 bacteriophages and therefore are of a completely different origin to the RNA polymerases of the eubacterial type.

How does transcription affect expression of the plastids' genes? Do differences in the level of transcription initiation control the amount of plastid transcripts, or is the amount of transcripts controlled mainly post-transcriptionally, by RNA stability? In order to answer this question, the equilibrium concentrations of the RNA need to be compared with the transcription rate and with the RNA degradation rate. Since synthesis and degradation rates of plastid transcripts in intact plants cannot be measured with the required degree of accuracy, systems in isolated organelles are used instead (Box **4.3** and Figure **4.14**).

In barley, plastid gene transcription rates can differ 300-fold (Table **4.5**). The transcription rate of individual genes is not constant, but can change as development progresses

Box 4.3 Measuring transcription activity using 'run-on' experiments

Run-on transcription is an experiment that is conducted in isolated organelles, and can be used to measure the *in vitro* transcription activity of single genes from a nucleus or organelle. In general, this technique can give a good representation of the conditions found *in vivo*. The run-on method requires intact organelles, and in the following example, they have been isolated by isopycnic centrifugation in a Percoll™ density gradient. Intact plastids are hypotonically lysed and radioactive nucleotides (usually [α-^{32}P]UTP) are introduced to the already initiated transcripts. After successful run-on transcription the transcripts are quantitatively hybridized with gene-specific DNA or RNA probes. Finally, the hybridized radioactivity level is measured. Experiments have shown that there is only very weak *de novo* initiation of transcription, meaning that the amount of radioactivity measured is a good representation of the level of transcription of a gene in a given tissue. To compare the transcription activity of specified genes, the lengths of the probes that were used has to be borne in mind, together with the frequency of the labelled nucleotide in the transcribed region that is illuminated by the probe.

from proplastid to chloroplast (Box **4.2**). For example, the proplastids in cells at the base of a barley leaf transcribe the *rpoB* gene at a relatively high rate. This rate decreases as development progresses. The genes *psbA* and *rbcL*, which are important for photosynthesis, behave in the opposite way: as the chloroplast differentiates their transcription rate increases, but in mature chloroplasts it decreases. This means that the transcription rates of plastid genes can be regulated in different ways.

Generally, a higher transcription rate is accompanied by greater amounts of transcripts. This is often reflected by the amount of proteins (although an exception to this is the accumulation of thylakoid proteins as etiolated seedlings turn green again; *see* section 5.6.5). At the least, in higher plants transcription is important during differentiation of chloroplasts and maintenance of their function.

However, a closer examination reveals that transcription rate, RNA, and protein amounts do not correlate precisely with one another. RNA stability (Table **4.5**) and translation (*see* section 5.6.5) are additional control mechanisms, and in some cases determine whether and how strongly a given gene is expressed. The light-dependent greening of etiolated barley seedlings provides a good example of translational control of plastid gene expression (*see* Figure **5.1**). Even though RNAs for the chlorophyll *a* apoproteins of photosystems I and II are present in the etioplast, these proteins are not synthesized, or are synthesized but are unstable and break down.

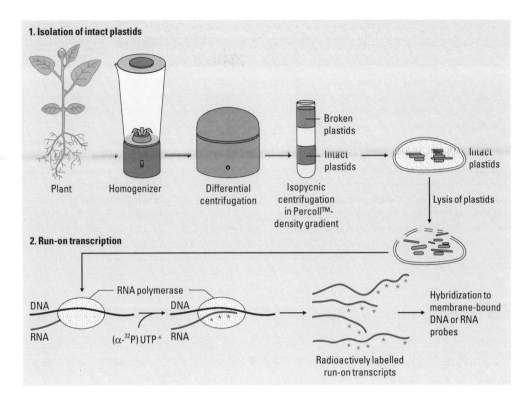

1. Isolation of intact plastids

Plant — Homogenizer — Differential centrifugation — Isopycnic centrifugation in Percoll™-density gradient

Broken plastids
Intact plastids
Intact plastids
Lysis of plastids

2. Run-on transcription

DNA

RNA polymerase

DNA

$(\alpha\text{-}^{32}P)\,UTP\,*$ RNA

RNA

Radioactively labelled run-on transcripts

Hybridization to membrane-bound DNA or RNA probes

Figure 4.14 Run-on transcription method. See Box **4.3** for explanation.

As with transcription rates, there can be large differences in the stability of plastid RNAs (*see* Table **4.5**). In any case, the stability of the RNAs is not constant and can change during development, as can transcription activity. Hairpin loop structures in the RNA participate in regulation of RNA stability. Such structures can be found at the 3' ends of many plastid transcripts (section 4.2.7). Meanwhile, preliminary experimental results suggest that, as with eubacteria, the 5' untranslated regions contain determinants for mRNA stability.

RNA equilibrium concentrations in plastids therefore always result from interactions between transcription rate and RNA stability (an example is the differential expression of plastid genes in mesophyll and bundle sheath chloroplasts of monocotyledonous C_4 plants—Box **4.4** and Figure **4.15**). Similarly, protein equilibrium concentrations are adjusted by synthesis and degradation. Therefore, the question is not which process is the most important for regulation of plastid gene expression, but instead how the processes interlink and whether they can be modulated individually or only together.

> ### Box 4.4 In the mesophyll and bundle sheath chloroplasts of the monocotyledonous C_4 plant *Sorghum bicolor*, plastid genes are differentially expressed (Figure 4.15)
>
> Monocotyledonous C_4 plants of the NADP malic enzyme type are characterized by a chloroplast dimorphism. Whereas the thylakoid membranes of the mesophyll chloroplasts are very similar to those of plants with C_3 metabolism, the grana in the bundle sheath chloroplasts are either not present, or are very rudimentary. A reduced number, or complete lack, of grana is correlated with a drastic reduction in photosystem II activity. By isolating intact chloroplasts of both types from *Sorghum bicolor*, it was shown that the different levels of *rbcL* and *psbA* transcripts in mesophyll and bundle sheath chloroplasts are accompanied by differences in transcription activity.
>
> However, differences found in the transcription levels of *rbcL* and *psbA* are insufficient by themselves to explain the observed levels of RNA. The different accumulation of *rbcL* and *psbA* RNAs in mesophyll and bundle sheath chloroplasts thus results from the interaction of different transcription activities and RNA stabilities.

Table 4.5 RNA amount, transcription rate, and the deduced RNA stability for selected plastid genes of barley. The measurements were made on isolated plastids from the apical leaf area of barley seedlings that had been grown in the dark for 4 days. The absolute RNA amount was determined by dot-blot hybridization. Synthetic transcripts of the examined genes were used for calibration. The transcription activity was estimated by the run-on transcription method. (According to Rapp, J. C., Baumgartner, B. J., and Mullet, J. (1992). *Journal of Biological Chemistry*, **267**, 21404–11).

Gene	RNA level	Transcription rate	Relative RNA stability
	fmol RNA	fmol UMP	RNA level
	5×10^6 plastids	5×10^6 plastids/ 5 min/kb	transcription rate
rRNA, tRNAs			
16S RNA	1183	98	12
trnfM-trnG	51.6	174	0.3
trnK-ORF 504	3.7	30.5	0.1
Photosynthesis			
rbcL	45.1	25.8	1.7
psbA	38.1	153	0.2
psbD	13.0	13.5	1.0
psaA	8.5	5.6	0.5
atpB	3.9	14.3	0.3
petB	12.5	4.3	2.8
NDH complex			
ndhA	0.3	2.4	0.1
Ribosomal proteins			
rpl16	2.4	2.4	1.0
RNA polymerase			
rpoA	1.6	1.2	1.3
rpoB	0.05	0.5	0.1

Taxa	Genome size (in kb)
Angiospermae	
Brassica hirta	208
Brassica campestris	218
Beta vulgaris	386
Cucumis melo	>2000
Zea mays (T-Cytoplasma)	540
Bryophyta	
Marchantia polymorpha	186
Chlorophyta	
Chlamydomonas reinhardtii	15.8
Prototheca wickerhamii	55.3

Table 4.6 Size and structure of selected chondriomes. The data are mainly taken from review articles by Palmer (1992) and Gray (1993). The genome size was estimated by restriction mapping, or sequencing of *Cucumis melo*. The estimation of the genome size is based on measurement of kinetic complexity.

4.3 The chondriome

4.3.1 The chondriomes of most land plants are composed of a dynamic population of DNA molecules

The size of chondriomes of land plants can be up to 2000 kb (Table **4.6**), and are much larger than the mitochondrial DNA of mammals, which is approximately 16 kb, and baker's yeast (*Saccharomyces cerevisiae*), which is approximately 80 kb. The structures of land plant chondriomes are also unusual. As a rule, the population of DNAs within the land plant chondriome is not as homogeneous as that of mammal chondriomes: it is a more or less complex population of DNAs with different sizes and sequences. *In situ* analyses have shown that in addition to the circular DNAs which only contain a part of the mitochondrial genome, there are also linear DNA molecules. Exceptions to this rule are the chondriomes of *Brassica hirta* and *Marchantia polymorpha*, which only contain a single circular DNA. A further exception is the green alga *Chlamydomonas reinhardtii* which has just a single linear DNA as its chondriome. Thus there is still a degree of controversy as to the exact structure of the plant chondriome *in vivo*. Compared to the plastome it is extraordinarily diverse and dynamic in its structure.

The extreme heterogeneity of land plant chondriomes is due to extensive recombination. The sequence repeats, which range in size from 50 to 1000 bp are possible explanations for this, as are the 1–14 kb duplications. Both sequence anomalies are typical of land plant mitochondrial DNA. Recombination even happens between different genes, so the chondriomes possess incomplete copies of various genes alongside intact ones. Recombination can

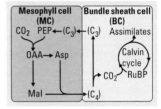

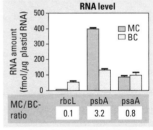

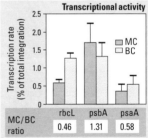

Figure 4.15 Differential gene activity in mesophyll and bundle sheath chloroplasts of C4 plants (*see* Box **4.4** for explanation). (From Kubicki, A. Steinmüller, K., and Westhoff, P. (1994). *Plant Molecular Biology*, **25**, 669–79.)

also create new chimeric genes. Such genes are thought to be responsible for male (pollen) sterility.

4.3.2 Coding capacity of plant chondriomes

The chondriomes of the land plants *A. thaliana* and *M. polymorpha* and the green algae *C. reinhardtii* and *Prototheka wickerhamii*, have been entirely sequenced. There is extensive mapping and incomplete sequence data available for a great variety of other mitochondrial DNAs. This enables scientists to compare the coding capacity of plant chondriomes. As with plastomes, the chondriome's coding capacity is not sufficient to code for all the known mitochondrial proteins and RNAs. For this reason, mitochondria, like plasmids, are genetically semi-autonomous.

About 94 substantially sized (60 amino acids or more) open reading frames can be located on the mitochondrial DNA of *M. polymorpha* (Table **4.7**). Therefore the coding capacity of this genome is much greater than its mammalian counterpart. Besides genes for the components of the translation apparatus and the respiratory chain, there are many open reading frames with unknown functions. Remarkably, the chondriomes of plants (especially angiosperms) are missing an entire set of chondriome-encoded tRNAs, so they have to import some tRNAs from the cytosol. The mechanism for doing this is unknown.

4.3.3 A brief look at mitochondrial gene expression

In comparison with the plastome of land plants, the chondriome gene order is only vaguely conserved. This is due to

Table 4.7 The gene repertoire of the *Marchantia* chondriomes. (According to Oda, K., Yamato, K., Ohta, E., Nakamura, Y., Takemura, M., Nozato, N., Akashi, K., Kanegae, T., Ogura, Y., Kochi, T., and Ohyama, K. (1992). *Journal of Molecular Biology*, **223**, 1–7.)

Function	Genes
Translation	
rRNAs	*rrn5, rrn18, rrn26*
tRNAs	29
Ribosomal proteins	*rpl2, rpl5, rpl6, rpl16, rps1, rps2, rps3, rps4, rps7, rps8, rps10, rps11, rps12, rps13, rps14, rps19*
Oxidative chain	
NADH dehydrogenase	*nad1, nad2, nad3, nad4, nad4l, nad5, nad6*
Cytochrome-*b/c* complex	*cob* (cytochrome-*b*)
Cytochrome-*c* oxidase	*cox1, cox2, cox3*
ATP synthase	*atp1, atp6, atp9*
Others	
Open reading frame	32

the extensive recombination events (section 4.3.1) that can very easily lead to new arrangements of the genes. Therefore the way mitochondrial genes are sorted into transcription units varies greatly within different plant groups. Many mitochondrial genes seem to be transcribed mono-cistronically, but there are also several examples of poly-cistronic transcription units. The plant mitochondrial RNA polymerase, like that of baker's yeast, is a bacteriophage-type enzyme which is encoded by the nuclear genome.

Mitochondrial genes are often interrupted by class I or II introns. While *Marchantia* still possesses both intron types, angiosperms have only class II introns. With class II introns, trans-splicing can often be seen, just as it can in class II introns of plastids. This means that intron sequences with certain 5' and 3' ends are split and are part of separated transcription units.

Analysis of sequences preceding the transcription initiation site show that they have nothing in common with prokaryotic promotor elements. However, homologous sequences can be found within monocotyledonous plants and within dicotyledonous plants, although there are none *between* the two types of plants. First studies on the *atp1* promoter of maize have shown that two domains in the region of nucleotide position −5/−7 and −11/−12 are essential for promoter function *in vitro*. However, further tests are necessary before drawing any general conclusions about the structure of mitochondrial promoters. Furthermore, how the expression of mitochondrial genes is regulated and whether mitochondrial genes can be differentially expressed are also unknown.

4.3.4 RNA editing in plant mitochondria and plastids: sequence differences between genes and their transcripts

If the sequences of mitochondrial genes and their corresponding mRNAs are compared, it is often the case that a cytosine in the gene will be replaced with a uracil in the corresponding mRNA, in a process called *RNA editing*. The reverse case also occurs, albeit much less frequently. RNA editing in plants is vital—only in this way can the correct mRNA sequence be made to act as a blueprint for successful protein synthesis. RNA editing is not only characteristic of coding regions, but also occurs in introns and even tRNAs. The details of the editing mechanism are poorly understood.

Editing of mitochondrial transcripts occurs in angiosperms, gymnosperms, ferns, and mosses, but has not yet been detected in the Charophyceae. RNA editing in plants is not limited to the mitochondria, but also occurs in the plastids of all major groups of land plants; however, it is far more commonplace in the mitochondria. As in the mitochondria, plastid RNA editing has not been observed in the Charophyceae, suggesting that RNA editing in both types of organelles has a common origin.

4.4 Interactions between different genetic compartments

The genetic material of plant cells has emerged from an endosymbiotic combination of three differently constructed subgenomes. As evolutionary development progressed, the two organelle genomes gradually lost most of their original genes, passing them into the nucleus. Finally, with the exception of genes for protein synthesis, and depending on organelle, genes for photosynthesis or the respiratory chain are left as functional genetic units which the plastome and chondriome can contribute to the plant cell. The considerable success of plants shows that they have managed to harness the three subgenomes together very successfully, and can co-ordinate their expression well. The following example of the interaction between plastome and genome will demonstrate which of the intergenomic regulatory mechanisms allow the output of the two subgenomes to be co-ordinated.

4.4.1 Interaction between genome and plastome, using *Oenothera* as an example

In the subgenus Euoenothera of the genus *Oenothera*, three genetically different nuclear genomes (A, B, and C) and five plastome types (I–V) can be distinguished. In nature, only certain plastome/genome combinations exist. Due to special circumstances (complex heterozygosity) during meiosis, only the terminal portions of the chromosomes undergo crossing over, so maternal and paternal genes do not really mix and are passed on to the offspring separately. In addition, the plastids are passed to offspring plants by both parent sexes, so by interspecific plant crosses all possible genome/plastome combinations can be created.

Not all the genome/plastome combinations can coexist in harmony; some combinations cause more disruption to the photosynthetic apparatus than others (Figure **4.16**). *Oenothera* provides an example of how, in closely related species, a highly complex interaction between plastome and genome can be so disruptive that recombination of the

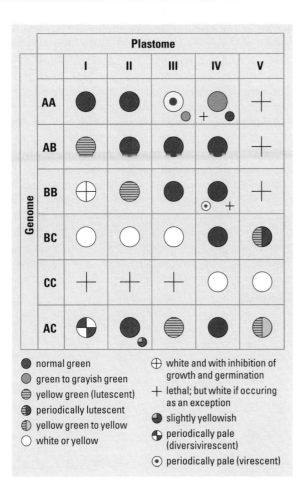

Figure 4.16 Genome–plastome incompatibility in *Oenothera*; for explanation see text. (Kindly provided by W. Stubbe, University of Düsseldorf, Germany.)

two subgenomes leads to massive dysfunction in the resulting plastids. Incompatibility between the nuclear genome and the plastome is actually a widespread phenomenon after interspecific crosses, and plant breeders refer to this outcome as 'hybrid bleaching'.

4.4.2 Nuclear genes control the expression of plastid genes, but so can plastids affect nuclear gene expression

Even though plastids possess a whole range of genes for photosynthesis, they cannot express them by themselves. Analysis of photosynthetic mutants has shown that nuclear genes are also necessary, which interfere at different levels with the expression of plastid genes. Nuclear genes that interfere with plastid gene expression do so

mainly at the post-transcriptional level. They are responsible for the correct processing of transcripts, regulate their
stability, and are required for the translation of mature
transcripts by the polysomes (Figure **4.17**). Even though, to
this date, genome-saturating mutagenesis experiments
have yet to be done in plants, it can be assumed, from similar experiments in yeast mitochondria, that the amount of
regulatory genes from the nucleus far outnumber those of
the plastids. Plastome gene expression is therefore under
the overall control of the nuclear genome.

Regulatory information also flows from plastid to nucleus. The functional state of the chloroplast affects those
nuclear genes that code for chloroplast proteins (for example, *CAB* for the chlorophyll-*a/b*-binding proteins of the
peripheral light-harvesting complex of photosystem II, and
rbcS for the small subunit of Rubisco). If chloroplast func-

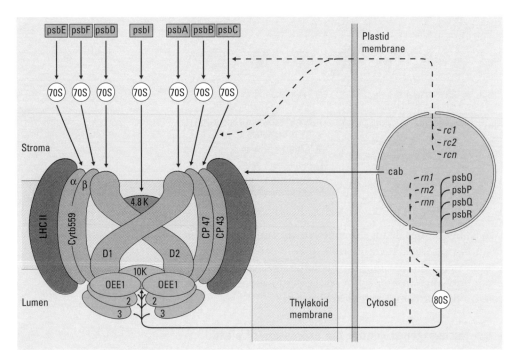

Figure 4.17 Control of plastid gene expression by
nuclear regulatory genes, for example photosystem
II biogenesis. The composition of photosystem II is
much simplified. The subunits D1, D2, cytochrome
$b559$, (Cyt $b559$ α and β) and the 4.8 kDa
polypeptide (4.8K) are the reaction centre.
OEE1/2/3 are the regulatory proteins of the water-
splitting complex. The inner antenna of
photosystem II consists of the chlorophyll proteins
CP47 and CP43. The peripheral antenna consists of
the various LHC proteins. The function of the
10 kDa protein (10K) is unknown. The regulatory
genes of photosystem II biogenesis are labelled as
rc1, *rcn*, and *rn1–rnn* (gene names for the
photosystem II subunits, *see* Figure **4.10**).
(According to Erickson, J. M., and Rochaix, J. D.
(1992). In *Topics in photosynthesis*, vol. 11, (ed.
J. Barber), pp. 101–77. Elsevier, Amsterdam.)

tion is halted because of a lack of carotenoids, the *CAB* and *rbcS* genes are switched off. Genes for cytosolic proteins, such as C_4 phosphoenol pyruvate carboxylase in *Zea mays*, are unaffected by this. Photo-oxidative disruption of chloroplast function affects only genes concerned with photosynthesis. In the case of the *CAB* genes, nuclear run-on experiments showed that the disappearance of the *CAB* mRNAs is regulated transcriptionally, not by RNA degradation.

The nature of the signal from the plastid, as well as the method of its transmission to the nucleus, is not known. However, in recent experiments on *Arabidopsis*, at least three different loci were identified that participate in signal transduction from plastid to nucleus. Signal movement is not one way: the plastid example demonstrates that over time two-way regulatory interactions have evolved between the three genetic systems of the cell. To fully understand this regulation, and how it can affect plant development, its mechanisms will have to be elucidated at the molecular level.

4.5 Bibliography

Nuclear genome

Abler, M. L. and Green, P. J. (1996). Control of mRNA stability in higher plants. *Plant Molecular Biology*, **32**, 63–78.
Good summary of effect of mRNA stability on plant gene regulation.

Benfey, P. N. and Chua, N.-H. (1990). The cauliflower mosaic virus 35S promoter: combinatorial regulation of transcription in plants. *Science*, **250**, 959–66.
Out of date review, but still useful background information.

Cohen, A. and Mayfield, S. P. (1997). Translational regulation of gene expression in plants. *Current Opinion in Biotechnology*, **8**, 189–94.
Up-to-date data on translational regulation in plants.

Dean, C. and Schmidt, R. (1995). Plant genomes: a current molecular description. *Annual Review of Plant Physiology and Plant Molecular Biology*, **46**, 395–418.
Current status of plant genome structure.

Doebley, J. (1993). Genetics, development and plant evolution. *Current Opinion in Genetics and Development*, **3**, 865–72.
Good description of the effects of transcription factors on plant evolution. Has information about genetic control of anthocyanin synthesis.

Gerasimova, T. I. and Corces, V. G. (1996). Boundary and insulator elements in chromosomes. *Current Opinion in Genetics and Development*, **6**, 185–92.
Good addition to Spiker and Thompson (1996). Detail about function of *Drosophila* chromatin boundary elements.

Martienssen, R. A. and Richards, E. J. (1995). DNA methylation in eukaryotes. *Current Opinion in Genetics and Development*, **5**, 234–42. Includes DNA methylation in plants.

Moore, G. (1995). Cereal genome evolution: pastural pursuits with 'Lego' genomes. *Current Opinion in Genetics and Development*, **5**, 717–24
Good addition to Dean and Schmidt (1995) with special emphasis on cereal genomes.

Ramachandran, S., Hiratsuka, K., and Chua, N.-H. (1994). Transcription factors in plant growth and development. *Current Opinion in Genetics and Development*, **4**, 642–6.
Good overview on plant transcription factors.

Spiker, S. and Thompson, W. F. (1996). Nuclear matrix attachment regions and transgene expression in plants. *Plant Physiology*, **110**, 15–21.
Summary of knowledge on matrix attachment regions in plants.

Plastome

Palmer, J. D. (1991). Plastid chromosomes: structure and evolution. In *Cell culture and somatic cell genetics of plants*, (ed. L. Bogorad and I. K.Vasil), Vol. VIIA, pp. 5–53. Academic Press, San Diego.
Good review.

Palmer, J. D. (1992). Comparison of chloroplast and mitochondrial genome evolution in plants. In *Cell organelles*, (ed. R. G. Herrmann), pp. 99–133. Springer, Wien.
Good addition to Palmer (1992).

Stern, D. B., Higgs, D. C., and Yang, J. J. D. (1997). Transcription and translation in chloroplasts. *Trends in Plant Science*, **2**, 308–15.
Good addition to Sugita and Sugiura (1996).

Sugita, M. and Sugiura, M. (1996). Regulation of gene expression in chloroplasts of higher plants. *Plant Molecular Biology*, **32**, 315–26.
Good starting point for this subject.

Chondriome

Gray, M. W. (1993). Origin and evolution of organelle genomes. *Current Opinion in Genetics and Development*, **3**, 884–90.
Still good summary of this subject.

Maier, R. M., Zeltz, P., Kössel, H., Bonnard, G., Gualberto, J. M., and Grienenberger, J. M. (1996). RNA editing in plant mitochondria and chloroplasts. *Plant Molecular Biology*, **32**, 343–65.
Extensive essay.

Schuster, W. and Brennicke, A. (1994). The plant mitochondrial genome: physical structure, information content, RNA editing, and gene migration to the nucleus. *Annual Review of Plant Physiology*, **45**, 61–78.
Good introduction.

Genome/plastome interactions

Barkan, A., Voelker, R., Mendel-Hartvig, J., Johnson, D., and Walker, M. (1995). Genetic analysis of chloroplast biogenesis in higher

plants. *Physiologia Plantarum*, **93**, 163–70.
Uses maize as an example.

Herrmann, R. G. (1997). Eukaryotism, towards a new interpretation.
In *Eukaryotism and symbiosis*, (ed. H. E. A. Schenk, R. G. Herrmann,
K. W. Jeon, N. E. Müller, and W. E. Schwemmler), pp. 73–118).
Springer, Heidelberg.
Long article, with exhaustive literature list, about evolution of
plant cell genomes and their interactions.

5 Light, phytohormones and the biological clock as inducers and modulators of development

Contents

5.1 **What is light and how does it affect the plant's life cycle?**

5.2 **Phytochrome as a prototype for plant photoreceptors**

5.2.1 The photoreceptor and its signalling chain; some general remarks

5.2.2 Phytochrome is a reversible red/far-red system

5.2.3 Phytochrome can turn genes on and off

5.2.4 Promoters of light-regulated genes have molecular 'light switches'

5.2.5 Phytochromes of higher plants are products of a gene family, and have specific physiological functions

5.2.6 The signal transduction chain of phytochrome—a cell biology approach

5.2.7 The *det* and *cop* mutants of *Arabidopsis thaliana*: a genetic approach to explain the light signal transduction chain

5.3 **Blue light—a system that induces developmental processes and protects against light stress**

5.3.1 CRY1—a blue light receptor of plants

5.3.2 Blue-light-regulated gene expression

5.4 **The biological clock**

5.4.1 Some basic observations

5.4.2 The biological clock drives gene expression

5.5 **Phytohormones**

5.5.1 The 'classical' phytohormones and their effects—a brief overview

5.5.2 The new phytohormones: jasmonates, oligosaccharides, brassinosteroids, and peptide hormones

5.5.3 In search of receptors and signal transduction chains

5.5.4 Phytohormones can switch genes on

5.6 **From the seed to the young seedling: phytohormones and light as regulators of development**

5.6.1 In angiosperms, seed development is divided into several phases

5.6.2 Searching for global regulators of late seed development

5.6.3 The role of abscisic acid in seed development

5.6.4 Germination

5.6.5 Early seedling development: from heterotrophy to autotrophy

5.7 **Bibliography**

Preface

A plant's development in consecutive phases does not follow a rigidly predefined plan. Its development is flexible, utilizing environmental information as well as its own internal information to decide how its development should proceed. Here, the plant's endogenous biological clock plays a vital part. External factors, such as light and temperature, can induce new development phases. Seed germination and flowering are probably the best-known examples of this. However, external factors often act only as modulators, affecting development quantitatively rather than qualitatively.

Light is the most important environmental factor, so we will examine it in some detail. To sense light, plants have evolved different kinds of photoreceptors, each with responsibility for different functions. We will become familiar with these photoreceptors, and how the cellular signal from these moves from the receptor to the target site (such as a gene to be activated within the nucleus).

Environmental factors such as light and temperature can act systemically, as well as locally at the level of the cell. To achieve this, plants have *signal molecules* that are able to transport internal and external signals around the plant body. Phytohormones are the plant's signal transporters, transmitting signals in the form of chemical messages between cells, tissues, and organs. Only recently, by the use of molecular and genetic techniques, have we begun to understand exactly how phytohormones work as signal transporters, and how they reach their cellular targets. Seed production and germination will provide us with two examples of central developmental programmes in flowering plants where phytohormones are key regulators. Both phases of development and the subsequent early formation of the seedling are good examples of how light can induce or influence developmental steps by interacting with phytohormones.

5.1 What is light and how does it affect the plant's life cycle?

Light is defined as 'that part of the electromagnetic spectrum that is visible to the naked eye'. It ranges between 350 and 700 nm, the spectral range between blue and red light. The eye is a very sensitive organ, but it is not an ideal instrument to measure light because it has a sensitivity bias towards green light at 550 nm; similarly intense light at

730 nm is perceived only weakly or not at all. A further complication is that light at 350 nm has twice the energy of light at 700 nm. For this reason, light is quantified as moles of photons (1 mol photons = 1 Einstein). To take into account the fact that different wavelengths have different energies, plant physiologists often refer to the *photon fluence rate* (mol photons/m^2/s), or the *photon fluence* (mol photons/m^2).

Because plants show physiological responses to electromagnetic radiation at wavelengths between 280 and 800 nm, the plant physiologist's definition of light differs slightly from the human biologist's. The spectrum of physiologically active wavelengths for plants extends into the UV-A and UV-B regions, and in the opposite direction into the infrared. As well as affecting photosynthesis, light affects the plant's development throughout its entire life cycle (Table **5.1**). This can be very obvious when two otherwise identical oak trees are compared. An oak tree that has grown in a densely populated forest, where it has had to compete for light with surrounding trees, will have grown very differently to one that grew alone in the middle of a meadow. There is no doubt that the varying amounts of light have modulated the growth of these trees, leading to extremely diverse shapes. Similar differences can be seen in weeds that grow either in the shade of other plants or out on their own. In every case, a plant's growth will differ according to whether neighbouring plants shade it or reflect light on to it. How the plant perceives light and uses it to modulate or induce developmental and adaptational processes is explained below.

5.2 Phytochrome as a prototype for plant photoreceptors

5.2.1 The photoreceptor and its signalling chain; some general remarks

For light to affect the plant, a signal must be delivered from the place where the light was intercepted, to the inner workings of the cells where the regulation mechanisms reside. To

Table 5.1 Phytochrome-driven developmental processes in the life cycle of an angiosperm plant.

Early phase
Induction of seed germination: light and dark germinators
Seedling development: scotomorphogenesis and photomorphogenesis

Adult plant
Avoiding shade, and perception of neighbouring plants
Induction of flower formation (photoperiodicity)

do this, the plant must be able to detect and analyse incident light for its intensity and wavelength; this it achieved by using photoreceptors that are characterized by their action spectra. Since the photoreceptor does not interact directly with the cell's DNA, there must be a signal transduction chain mediating between the two. Elucidation of such signal transduction chains is still in its early days, and they are much better characterized in more extensively studied animal systems. At least three types of light receptors can be distinguished: phytochrome, blue light/UV-A, and UV-B receptors. It is rather striking that the 'mass-manufactured' pigments, the chlorophylls and carotenoids, are not used to sense light quality and quantity and that instead specialized receptor pigments have developed. It is also important that these receptors are capable of measuring the extreme blue and red ends of the light spectrum.

We have to use a model system to aid our understanding of the effects of light at the molecular level. Molecular biologists and biochemists take a rather pragmatic approach to choosing a suitable model system—they tend to pick molecular processes that give striking quantitative and qualitative changes within a short time span.

From this perspective, etiolated seedlings are ideal model systems for studying the effects of light. If seedlings are germinated in the absence of light, the plant adopts a characteristic growth habit, such as having long internodal lengths and small, rudimentary leaves. These growth habits are described by the term *etiolation*. Anyone who has seen what a potato that was germinated in the dark looks like will be well acquainted with this phenotype. What makes this ideal as a model system is that etiolated plants revert to normal growth very quickly when they are placed in the light. Developmental changes that are regulated by light are called *photomorphogenic*, and those that happen in the dark are called *scotomorphogenic* (Figure 5.1 and Table 5.2).

An excellent example of photomorphogenesis can be seen in the Gramineae. In many grasses (such as barley), the primary leaf is as tall in an etiolated plant as in a light-grown plant of the same age (Figure 5.1). When investigating the effect of light this is actually an advantage, because cell division and cell elongation do not interfere with the greening of the plant. Because chlorophyll synthesis depends on light, the primary leaf of an etiolated seedling is white or yellow because of the presence of lutein. After 24 hours of constant light the leaf will become completely green and its etioplasts are transformed into photosynthet-

Table 5.2 Developmental strategies of *Arabidopsis thaliana* in the dark (scotomorphogenesis) and in the light (photomorphogenesis).

Scotomorphogenesis
Long hypocotyls
Closed apical hooks
Small rudimentary cotyledons
Rudimentary leaf differentiation, for example no stomata
Etioplasts
Photomorphogenesis
Short hypocotyl
Open hypocotyl hooks
Expanded cotyledons
Differentiated cell types in the leaf
Functional chloroplasts

Figure 5.1

Photomorphogenesis in monocots (example: barley) and dicots (example: pea). The length of the pea seedlings grown in light or dark differs greatly. The leaves do not develop in the dark and the hypocotyl hook remains closed. If germination and growth start in the light, the leaves are well developed. The distances between the leaves and lengths of the internodes are rather short compared to those of an etiolated plant. In barley there is hardly any difference in the length of the primary leaves, regardless of whether the plant is etiolated. However, the coleoptiles will be about twice as long if the plant has been grown in the dark. To show this dramatic effect, in each case one of the primary leaves has been removed from the coleoptiles.

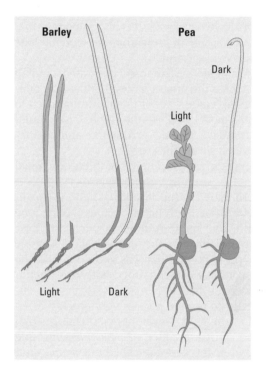

ically active chloroplasts (*see* Figure **4.9** and Box **4.2**). This is a dramatic transformation, and it leads to the expectation that on the molecular level a massive *de novo* synthesis occurs (Figure **5.2** and Box **5.1**). But which photoreceptors transmit the light signal and measure the light intensity?

5.2.2 Phytochrome is a reversible red/far-red system

The phytochrome system is the best understood of the plant photoreceptors. Its concept is a brilliant one. Phytochromes

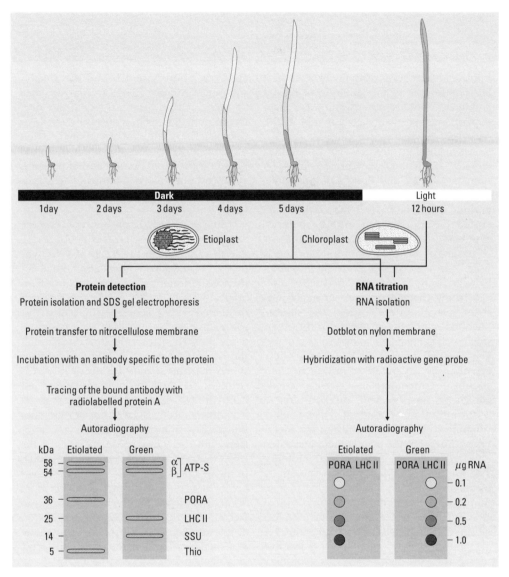

have two possible states; these differ in their absorption maxima (660 and 730 nm) and phytochrome can switch from one state to the other when it intercepts light at its current absorption maximum (for example, light at a wavelength of 660 nm will cause the 660 nm form to change to the 730 nm form, *see* Figures **5.3** and **5.4**). Phytochromes are chromoproteins composed of an apoprotein of about 120 kDa and a covalently linked chromophore, which is a linear tetrapyrrole system; phytochromes act as dimers (Figure **5.5**). In general, P_{fr} (that is, the 730 nm-absorbing

Figure 5.2 Change in protein and mRNA levels during greening of etiolated barley seeds. For explanations, see Box **5.1**.

Box 5.1 The greening of etiolated barley seeds: a model system with which to study light-regulated gene expression (Figure 5.2)

Etiolated barley seeds are a favourite material for studying light regulation of plastid and nuclear photosynthesis genes. For these experiments, seeds are germinated in the dark at 25 °C, and the seedlings are kept for 5 days in absolute darkness. The fully differentiated cells of the leaves form etioplasts by then (*see* Box **4.2** and Figure **4.1**). After this, plants are transferred to the light, and at various time points, leaf material is harvested for protein and mRNA analyses. Nowadays, for specific quantitative analysis of particular proteins, immunoblotting is used. RNA concentrations can be measured by hybridization with gene-specific probes.

To show changes of levels of single proteins using immunoblotting, the soluble membrane proteins are first split into their single polypeptide chains by use of a detergent (sodium dodecyl sulphate, SDS). These can then be separated according to their sizes by electrophoresis on a polyacrylamide gel. The separated proteins are transferred to a nitrocellulose membrane using electrophoresis. They bind tightly to the membrane. After this, the membrane is incubated with antibodies against particular proteins. The bound antibodies can be shown by incubating them with radiolabelled protein A, and subsequent autoradiography. Protein A is a cell wall protein of *Staphylococcus aureus* that can be used as a specific probe for γ-immunoglobulin antibodies. It is possible to see that the light-harvesting complex

(LHC) of photosystem II and the smaller subunit of Rubisco (SSU) are up-regulated by light. Other proteins, such as protochlorophyllide oxidoreductase (PORA) or thionin (Thio), are down-regulated. Levels of these proteins decrease upon illumination. In comparison, levels of other proteins, such as the α- and β-subunits of ATP synthase (ATP-S-α/β) are almost unaffected by light.

Using gene-specific probes, one can see whether the differing protein amounts are due to variations in the levels of RNA synthesis. Isolated RNA is applied as a small dot (hence the term, *dot blot*) to a nylon membrane, to which it adheres. The membrane is incubated with a gene-specific radioactive probe, and can be visualized by exposing to X-ray film. The darker the dots on the film, the more RNA there is. The experiment shows that the amount of LHC-II mRNA is strongly elevated after illumination, but PORA mRNA decreases drastically. Run-on experiments using isolated nuclei show that the decrease or increase in RNA levels correlates with the transcription activities of the genes (for methods, *see* Box **4.3** and Figure **4.14**). In higher plants, the light-dependent regulation of these nuclear-encoded genes therefore occurs mainly at the level of transcription (section 5.2.3). In contrast, the production of plastome-encoded proteins is usually regulated post-transcriptionally (for instance, in barley; *see* sections 4.2.8 and 5.6.5).

---

far-red form) is considered to be the active state because many photomorphogenic processes are induced by red light. However, it must be recognized that because of overlapping absorption spectra, both states exist in a photodynamic equilibrium. Seeing phytochrome as a molecular switch might therefore be more appropriate. Using an external factor—light—this switch enables the plant to go down a particular developmental road, without having to define which state of phytochrome causes which particular effect (Figure **5.4**).

Phytochrome-dependent reactions have been classified in terms of photon fluence and timing of irradiation, into two categories. These are the *induction reaction* and the *high irradiance reaction* (HIR). The induction reaction is further subdivided into the *low fluence reaction* (LFR) and the *very low fluence reaction* (VLFR). The low fluence reaction is the classic phytochrome-dependent response. It is characterized by photo-reversibility. A pulse of far-red light immediately following a red pulse will reverse the phytochrome's

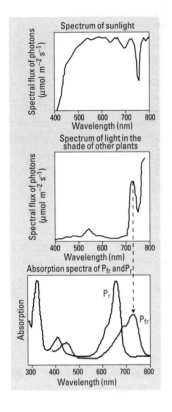

Figure 5.3 The importance of the phytochrome system in measuring light quality during the development of a plant in the shade. The function of the phytochrome system is not only linked to the early stages of seedling development, but also senses the quality of light in the adult plant. This measuring system is important for plants that grow in the shade of other plants or close to competing neighbouring plants. It delivers information at a particular time point when there is no direct shade from a neighbouring plant. The light spectrum is shown in the upper part of the figure. Below 400 nm there is comparatively little light energy and the decrease in light energy at 760 nm is probably because of the water content of the atmosphere. The light that passes through the plant cover has a characteristic spectrum (see middle diagram). Besides a low amount of green light (green gap) the light is mainly at a wavelength higher than 700 nm. The arrow shows that P_{fr} is excited by this light and the equilibrium is shifted to P_r. The light reflected from green plants has a similar composition. In this way it is a signal for the plant not to overgrow because P_r is formed in the side leaves. This phenomenon can explain the edge effect in fields: plants in the middle of a field grow more uniformly and higher than those on the edge.

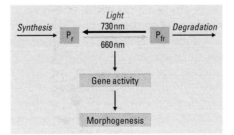

Figure 5.4 A simple model of the phytochrome system. The model does not account for the plant's possession of several phytochromes with different action spectra (*see* section 5.2.5).

Figure 5.5 Domain structure of phytochromes. The amino acid sequences of several dicotyledonous and monocotyledonous type A phytochromes (*see* section 5.2.5), together with different experimental results, form the basis for this model. The chromophore (dark red box, domains C and D) as well as the centres of positive (+) and negative (−) charge are depicted. (According to Furuya, M. and Song, P. S. (1994). In *Photomorphogenesis in plants* (ed. R. E. Kendrick and G.H.M. Kronenberg), pp. 105–40. Kluwer, Dordrecht.)

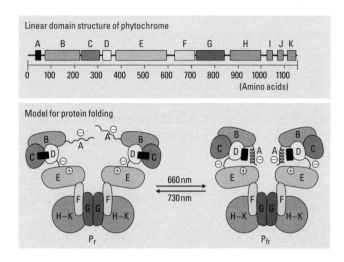

effect after the red pulse. This gives us a criterion by which to decide whether a given light effect happens because of a phytochrome. The low fluence reaction requires a photon fluence of at least 1 μmol/m^2 of red light. This accounts for a share of 2 per cent of the P_{fr} form of the total amount of phytochrome. The amount of time permitted before the effect becomes irreversible between the first red and second far-red light pulses depends on the particular effect that is being investigated.

The very low fluence reaction needs less than 0.1 nmol/m^2 of red light. This equates to only 0.01 per cent of total phytochrome being in the P_{fr} state. A very low fluence response can also be achieved using very short-duration far-red light pulses, and this is irreversible.

The high irradiance reaction shows peak response in the far-red and blue areas of the spectrum. As its name suggests, this response depends on the flow of energy. It can usually only be seen in etiolated seedlings that are kept permanently in far-red light. The response is irreversible.

Under laboratory conditions all three reaction patterns of phytochromes can be distinguished. However, in nature they often act in concert, regulating many different processes. The germination of *Arabidopsis* seeds (Figure **5.6**) is a good example. At first glance, the induction of seed germination can be described as a photoreversible low fluence reaction. However, on closer examination, especially of phytochrome mutants, one can see that the seed germination can also be induced by a very low fluence reaction. In section 5.2.5, we will see that the different types of reactions of the 'phytochromes' are due to bio-

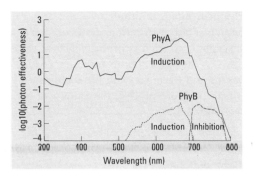

Figure 5.6 Action spectra for the light induction of seed germination in *PHYA* (hy8) and *PHYB* (hy3) mutants of *A. thaliana*. The action spectrum of phytochrome A for the induction (closed line) has been measured in the *PHYB* mutant. The induction of germination cannot be reversed—it is a very low fluence reaction. The action spectrum of phytochrome B for induction (dotted line) and inhibition (interrupted line) has been measured in *PHYA* mutants. They show the typical photoreversibility of low fluence responses. The photon efficiencies (50 per cent germination) were corrected according to the transparency of the seed testa for the specific light qualities. (According to Furuya and Schäfer 1996.)

chemically and physiologically different species of these molecules.

5.2.3 Phytochrome can turn genes on and off

In 1979 it was first reported that not only white light, but also a short red (660 nm) pulse, can induce the accumulation of LHC-II mRNAs in etiolated barley seedlings. The stimulatory effect of the red light was reversed by far-red (730 nm) light. This showed that LHC-II mRNA accumulation in etiolated chloroplasts seedlings is controlled by the phytochrome system. Several phytochrome-induced mRNAs are now known. To this group belong mRNAs for components of the photosynthesis complexes, and mRNAs required for producing enzymes for anthocyanin metabolism (chalcone synthase) and for nitrate assimilation (nitrate and nitrite reductases). The speed with which the effects of phytochrome are manifested at the mRNA level leads to the conclusion that the signal transduction chain responsible is relatively short, and that no other light-induced processes are included.

Regulation of phytochrome-driven gene expression is, at least in higher plants, at the level of transcription. This was shown by run-on transcription experiments with isolated nuclei from etiolated and subsequently red-light-irradiated seedlings (for the methods that were used, *see* Box **4.3** and Figure **4.14**). If the isolated nuclei were incubated with radiolabelled ribonucleotide triphosphates (rNTPs), the labelled rNTPs were integrated into transcripts that were already initiated. The labelled transcripts were quantified by hybridization to gene-specific probes. Using this technique, it was shown that phytochrome-induced increases in the level of LHC-II mRNAs were due to transcription, and not to any enhanced stability of the LHC transcripts.

Phytochrome is multitalented, not only can it switch genes on, it can also turn certain genes off. An example of mRNAs susceptible to such *down-regulation* of transcription are the mRNAs for protochlorophyllide oxidoreductase A (*see* Box **5.1** and Figure **5.2**) and phytochrome's very own mRNA (or, to be more precise, the mRNA of phytochrome A; *see* section 5.2.5).

5.2.4 Promoters of light-regulated genes have molecular 'light switches'

The molecular basis of light-driven regulation at the transcription level is only partially understood. *Cis*-regulatory sequences (molecular switches) were discovered following experiments in which deletions (for the methods, *see* section 4.1 and Figure **4.5**) were made in promoters of light-regulated genes, which were known to be indispensable. The promoter of the pea *rbcS-3A* gene has the sequence motif GTGTGGTTAATATG. This is called Box II, and it binds a transcription factor called GT-1 (Table **4.2**). If this sequence is transferred as a quadruplet into a heterologous light-insensitive promoter, the latter becomes light-regulatable. Also, the promoter of the chalcone synthase gene of parsley (Figure **5.7**) has two sequence motifs, I and II, which make it light-regulatable. If *unit 1* is fused to the TATA box region of the 35S promoter (*see* Figure **4.6**), this hybrid becomes light-regulatable, too. Sequence motif II of unit 1 binds to a transcription factor of the bZIP type. Sequence element I has a binding site for a Myb protein (*see* Table **4.2**).

Currently, it is not known whether the sequence motifs mentioned in the previous paragraph are the sole 'light switches', or whether there is a variety of *cis*-regulatory elements that are different in each gene. It is also not known whether the different photoreceptors use the same light switches, or whether the plant has a different sequence motif and transcription factor for every photoreceptor. It must also be clarified whether other signal chains (for instance, ones for organ specificity) interfere with these molecular light switches.

As well as sequences that are vital if a promoter is to be light-regulatable, one can also find other sequence elements that determine the strength of the response. But, by themselves, these sequences give no up- or down-regulation response to a light pulse. These sequence elements can be arranged as imperfect repeats, and according to the number of these, the strength of the response is determined.

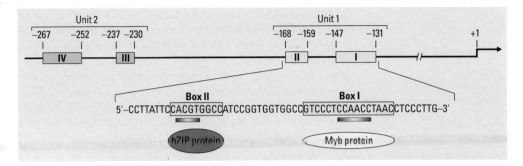

Figure 5.7 Molecular light switch of the chalcone synthase promoter of parsley. The promoter contains four sequence motifs, Box I to IV, that participate in light regulation. Their position in regard to the transcription starting point (+1) are shown (in base pairs, bp). White light containing UV is most effective. Blue and red light only modulate the effect. As explained in section 5.2.4, unit 1 can make a heterologous promoter light-regulatable. Unit 2 is therefore more or less redundant. The interaction of a transcription factor of the bZIP type (Table **4.2**) with Box II has been shown experimentally. The central motif of Box II (ACGT) is highlighted. Box I contains a binding motif (CCAACC) for a Myb protein (Table **4.2**). (According to Weisshaar, B., Armstrong, G. A., Block, A., Da Costa e Silva, O., and Hahlbrock, K. (1991). *The EMBO Journal*, **10**, 1777–86.)

5.2.5 Phytochromes of higher plants are products of a gene family, and have specific physiological functions

In the 1970s and 1980s, physiological and spectroscopic experiments revealed that there are, in fact, two different pools of phytochromes in the plant. In etiolated plants there is a pool of labile phytochrome (phytochrome I) that decreases rapidly when the plant is exposed to light. Together with this, there is a pool of stable phytochrome (phytochrome II), and the amount of this is unaffected by light. Immunological and protein chemistry experiments showed that the two pools have different phytochrome proteins, called Type I or II. Only molecular biology, by sequencing the genes, was able to reveal that there are several phytochromes in a plant, coded for by different genes. In *Arabidopsis thaliana* the phytochrome gene family consists of five genes, named *PHYA* to *PHYE*. *PHYA* genes have been isolated from various monocotyledonous and dicotyledonous plants. They resemble each other (their amino acid sequences are 65–80 per cent identical) more than other members of the *PHY* gene family. This might be because they share a common evolutionary origin.

The *PHYA* genes code for Type I phytochrome (the one that accumulates in etiolated seedlings and is quickly degraded in the light). The transcription of the *PHYA* genes is regulated by phytochromes—they are switched off in the light (Figure **5.8**). Type II phytochrome seems mainly to be a product of *PHYB*. The amounts of protein products of the other genes from this pool have yet to be evaluated. *PHYB* and *PHYC* are constitutively expressed, and the amount of their mRNAs is not influenced by light (Figure **5.8**).

The first insights into the functions of particular phytochrome types were from investigations of phytochrome mutants. They come from the group of so-called *hy* mutants. This group contains all the mutants that, even when grown in the light, have the typical phenotype of a dark-grown plant: an elongated hypocotyl.

The *Arabidopsis* mutant *hy8* completely lacks phytochrome A. This doesn't seem to cause any morphological effects if the seedlings are grown in white light, or constant red light at 660 nm. Like the wild type, *hy8* seedlings develop small hypocotyls under these conditions. Phytochrome A therefore seems not to be important for the plant's development in white light. The case is different, though, if the plant is grown in constant far-red light—*hy8* plants develop long hypocotyls, but in this light hypocotyls of the wild types stay short. A short hypocotyl in constant far-red light is typical of the high irradiance response of phytochromes. Therefore, without phytochrome A, there is no high irradiance response.

However, phytochrome A is also the photoreceptor for the very low fluence reaction. If wild-type *Arabidopsis* seeds are soaked in the dark for 48 hours they need a much lower photon fluence to induce germination than if they had not been pre-treated this way. However, the *hy8* mutant has lost sensitivity to very low photon fluences. The germination of *hy8* seeds will only occur via the

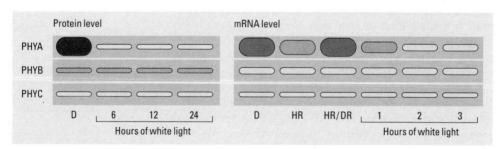

Figure 5.8 Regulation of the phytochrome gene family in *Arabidopsis thaliana*. Of the five phytochrome genes (*PHYA* to *PHYE*) *PHYA* to *PHYC* have been studied intensively at the RNA and protein levels. In etiolated plants (D) phytochrome A is the most abundant phytochrome. Its ratio to phytochrome B and C is 500 : 10 : 1. If the plants are irradiated with red light (HR) (or white light), phytochrome A levels decrease drastically. The phytochrome A : B : C ratio is 5 : 10 : 1 after 24 hours of constant light. The levels of the *PHYA*

mRNA transcripts (but not of those of *PHYB* and *PHYC*) decrease in red light (or white light). The reduction is much greater in monocotyledonous plants (barley, oats, and rice) than in dicots. The red light effect can be reversed by far-red light (DR). The *PHYA* gene regulates itself (autoregulation). In oats it has been shown that the reduction of *PHYA* mRNA levels is controlled by transcription. The promotor of the gene contains a *cis*-active sequence element which is essential for autoregulation. (According to Quail 1994.)

photoreversible low fluence reaction of the phytochrome (Figure **5.6**). Phytochrome A is therefore an important photoreceptor for seed germination and early seedling development. The *hy3* mutant of *Arabidopsis* is phytochrome B deficient. Seeds of this mutant cannot be induced to germinate by the photoreversible low fluence reaction of the phytochrome. Phytochrome B is therefore the photoreceptor of the low fluence reaction. However, the early seedling development of *hy3* plants is also affected. Phytochrome B mutants such as *hy3* have elongated hypocotyls if grown in white light and constant red light, but not in constant far-red light. Therefore, their behaviour is exactly the opposite of the phytochrome A mutants. Besides having long hypocotyls, they also have reduced cotyledon growth and chlorophyll pigmentation. This shows that phytochrome B plays an important role in germination and seedling development. Adult *hy3* plants look like seedlings, with their elongated hypocotyls and petioles, and smaller leaf blades; they thus behave as though constantly shaded by other plants, and have a constant *shade avoidance syndrome*. These plants also flower much earlier than the wild type. Phytochrome B is therefore important for regulating the growth and development of adult plants according to their environment.

5.2.6 The signal transduction chain of phytochrome—a cell biology approach

Experiments in yeasts and lower and higher animals have shown that signal transduction mechanisms are conserved in eukaryotes. Surprisingly enough, it was found recently that there are also similarities between prokaryotes and eukaryotes. For these reasons, we assume that plants should also possess typical components of signal transduction chains, such as G proteins, phosphatidylinositol phosphate, Ca^{2+}/calmodulin, and protein kinases of differing specificities. Recently, these assumptions have been increasingly confirmed. Using the findings from the animal world, it seems feasible to construct a model system for the signal transduction of light in plants. A cell biology approach neatly demonstrates the strengths of this model system, discussed below.

The starting point for these cell biology investigations was the *aurea* mutant of tomato. Its etiolated seedlings have only 5 per cent phytochrome A, much less than in the wild type. Processes in green plants that are regulated by phytochrome A are also affected: *aurea* seedlings have only

20 per cent of the normal wild-type anthocyanin content, chlorophyll is greatly reduced, and the thylakoid membranes of the leaf chloroplasts have no grana. When etiolated *aurea* seedlings were placed in the light, no chloroplasts developed in their hypocotyl cells, and no anthocyanin accumulated. Microinjection of phytochrome A led to the development of chloroplasts and an accumulation of anthocyanins. Reversion experiments with red and far-red light confirmed that both effects depended on the presence of P_{fr}. This gave researchers a test system where they could use just a single cell and the microinjection technique to identify possible intermediates in the phytochrome signal transduction chain. The test system was further refined later, when co-injection of promoter/ reporter gene constructs was used to measure direct effects at the level of the genes.

Figure **5.9** shows the results of these experiments. The central molecule following phytochrome A is a heterodimeric G protein. After this the signal transduction chain splits into two separate paths. A Ca^{2+}/calmodulin-dependent path leads to the activation of the CAB (chlorophyll-*a*/*b*-binding protein) promoter. Cyclic GMP (cGMP) transmits the signal to the CHS (chalcone synthase) promoter, while both cGMP and Ca^{2+}/calmodulin are required to activate the FNR ($NADP^+$-ferredoxin oxidoreductase) promoter. These experiments led to an experimentally verifiable model for the signal transduction chain of phytochrome A, and provided the basis for further similar experiments.

5.2.7 The *det* and *cop* mutants of *Arabidopsis thaliana*: a genetic approach to explain the light signal transduction chain

In order to analyse a signal transduction chain using a genetic approach, one usually looks first for mutants that show a response without having to be stimulated. This implies that the signal transduction chain contains elements that suppress a response in the absence of any stimulus. The stimulus itself abolishes the negative effects of the suppressing elements. A constitutive phenotype can be made by mutating suppresser elements in the signal transduction chain.

As was discussed earlier, depending on the presence or absence of light, plants have two different developmental strategies: scotomorphogenesis or photomorphogenesis (Table **5.2**). On this basis, a very simple selection method for

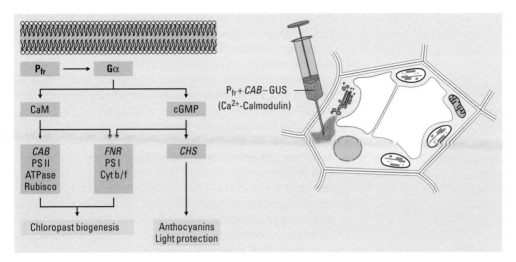

Figure 5.9 A model for the signal transduction chain of phytochrome. As a gene marker for anthocyanin synthesis, the chalcone synthase promoter (*CHS*; *see* section 4.1.5 and Figure **4.7**) was used. The promoters of LHC-II (*CAB*) and the NADP⁺-ferredoxin oxidoreductase gene (*FNR*) were used as representative photosystem II- or photosystem I-associated genes. The subunits of the two photosystems, the cytochrome-b_6f complex (cyt *b/f*), the ATP synthase (ATPase), and Rubisco have been elucidated immunocytochemically. Starting from P_{fr}, which, according to these assumptions, is at least temporarily associated at the plasma membrane, the α-subunit of a heterotrimer G protein (Gα) is activated. This leads, on the one hand, to an increase in intracellular Ca^{2+} concentration and, on the other hand, to a stimulation of guanylate cyclase activity. Calmodulin (CaM) induces *CAB* transcription but also activates other PSII genes, as well as the genes for the small subunit of Rubisco (*rbcS*) and for the subunits of ATP synthase. cGMP on its own activates the secondary metabolism that leads to the synthesis of anthocyanins. Switching on of the *FNR* promoters, as well as the genes for the subunits of PSI and the cytochrome-b_6f complex, is cGMP and Ca^{2+}/calmodulin-dependent. The biogenesis of functional chloroplasts therefore requires a simultaneous activation of both regulatory pathways. Meanwhile, there are increasing hints that the cGMP and Ca^{2+}/calmodulin pathways influence each other negatively. (According to Barnes *et al.* 1997.)

constitutive photomorphogenic mutants can be devised. Amongst mutant plants that have been grown in the dark, one searches for plants with phenotypes as if they had been grown in the light; in other words, those with short hypocotyls. Using this strategy, several mutants of *Arabidopsis thaliana* have been isolated, called *det* ('de-etiolated') or *cop* (*constitutively photomorphogenic*). All *det*/*cop* mutants show pleiotropic effects as seedlings grown in either dark or light. These effects also include photomorphogenic features. Also belonging to the constitutively photomorphogenic mutants are most of the *fusca* mutants (so called because their seeds appear dark due to the accumulation of anthocyanins; *fuscus* is Latin, meaning dark, black). Complementation analyses show that most of the *DET/COP* genes are allelic to the *FUSCA* loci.

DET/COP loci that possess mutant alleles of the *fusca* phenotype are essential for the complete expression of the scotomorphogenic programme in dark germination (Table **5.2**). Mutants of these loci lead to short hypocotyls in dark-germinated seedlings and expanding cotyledons. The plastids differentiate in the dark into chloroplast-like forms and, unlike in the wild type, into etioplasts. Transcripts of nuclear and plastid photosynthesis genes (for example *rbcS, cab, psbA*) that are induced in the wild type by light accumulate in the *det/cop* seedlings in the dark. The simplest explanation for this is that wild-type *det/cop* alleles suppress photomorphogenesis when plants are grown in the dark. This means they act as repressors, and are inactivated once the plant is put in the light. If the *det/cop* mutants are combined as double mutants with mutations also to the phytochrome system (for example, *hy1, hy2*, and *hy3*) or to the blue light receptor, *hy4*, the phenotype of the corresponding *det/ cop* mutant is always expressed. This means that both the signal transduction chain of phytochromes and the blue light receptor use the *DET/COP* gene products. The *DET* and *COP* gene products are therefore not unique to the signal transduction chain of phytochrome.

The primary structures of DET1, COP1, COP9, and FUS6 are known, and initial functional analyses have been carried out. The COP1 protein is present in the nucleus of dark-germinated seedlings and is transported into the cytosol upon illumination. Therefore it acts as would be expected for a repressor of gene expression. Also, the *DET1, COP9*, and *FUS6* gene products are nuclear proteins, but they are not transported into the cytosol upon illumination. COP9 and FUS6 proteins are parts of a high molecular weight complex which contains other unknown proteins. This complex is not assembled in mutants. The COP8 and COP11 proteins are therefore required for the assembly of the COP9 and FUS6 complexes. How these complexes act is unknown.

The *det/cop* mutants have been isolated because in the dark they display light phenotypes. Researchers hoped that in this way they would be able to isolate parts of the light signalling chain. However, these expectations have only been partly fulfilled. Even though investigations to date do not allow any final conclusions to be drawn, one can already infer that the DET/COP proteins are not parts of the phytochrome signal transduction chain, nor that of the blue light receptor. The *DET/COP* genes seem to be a general gene repression system that is essential for the transition between scoto- and photomorphogenesis. However, new investigations have shown that they are also used in other processes as negative regulators of gene expression. The fact that homologous proteins have been found in mammals also reinforces the idea that the DET/COP proteins have other, more generalized functions.

5.3 Blue light—a system that induces developmental processes and protects against light stress

That blue light has a dramatic effect on morphogenesis has been known for a long time. Short-term effects of blue light have been observed in phototropism and stomatal closure. Under constant red light, fern prothalli develop into a thread-like structure and the alga *Acetabularia* does not form a cap structure. Only if blue light is provided does the prothallium develop into a thallose shape and the *Acetabularia* cell form a cap. In a similar fashion, *Arabidopsis* calluses grown in red light do not regenerate roots, shoots, or flowers. Formation of these organs and induction of flower development all depends on blue light. Neither the mechanisms of the signal transduction chain involved nor the morphogenesis itself are understood. There are good reasons for believing that the UV and blue light receptors are old in evolutionary terms. They seem to have developed in an era when UV irradiation was much stronger than it is today. The available data further suggest that there are several types of blue light receptors that act in parallel, although independently. In many cases, extremely low light levels are sufficient to trigger blue light effects.

The blue and UV radiations are more energetic than red light, and so more harmful to the organism. Indeed, the energy of UV-A and UV-B is sufficient to cause mutations (UV-C barely reaches the Earth's surface).

Definition UV-A = 320–400 nm, UV-B = 280–320 nm, UV-C = 200–280 nm.

This means that blue light receptors also help the plant to sense dangerous radiation. As an example, many less-pigmented organisms, such as fungi, react to blue light by producing carotenoids. These help them to filter and absorb dangerous blue light. Similar systems can be seen in higher plants.

5.3.1 CRY1—a blue light receptor of plants

Although they have been extensively studied physiologically and biochemically, until recently the nature of blue light receptors was largely a mystery. Their action spectra indicate that flavins, carotenoids, or cytochromes may make up the receptors. Unfortunately, since these substances are all parts of cellular reaction chains, their action spectra tell us little (in comparison to those of phytochromes). The blue light receptors have been called *cryptochromes* (*crypto* = Greek, 'hidden'), a reasonable name in light of their somewhat mysterious nature. However, recently there was a major breakthrough in the search for blue light receptors.

One of the *hy* mutations (section 5.2.5), *hy4*, shows no blue light-dependent inhibition of hypocotyl growth; the phytochrome responses of *hy4* plants remain similar to the wild type. This is a strong hint that a component in the blue light signal transduction chain or the blue light receptor itself is defective in this mutant. Its gene (*HY4*) has been isolated and molecular, biochemical, and physiological analyses have verified that it encodes a blue light receptor. For this reason the gene was renamed as *CRYPTOCHROME1* (*CRY1*). The amino-terminal end of the CRY1 protein has similarities to microbial photolyases. The carboxy-terminal end is similar to the muscular protein tropomyosin (Figure **5.10**). Photolyases are DNA repair enzymes which split pyrimidine dimers that have been formed by the action of UV irradiation. They bind a deazaflavin or pterin molecule non-covalently to their amino terminus. These chromophore groups collect light energy and transfer electrons to a flavin dinucleotide (FAD) at its carboxy terminus. Finally the cyclobutane ring of the pyrimidine dimer is split by reduction.

The *CRY1* cDNA can be expressed in *E. coli* or insect cells to obtain a functional protein. The recombinant CRY1 protein binds FAD and a pterin non-covalently. As predicted it has no photolyase activity. An overexpression of the CRY1 protein in tobacco or *Arabidopsis* results in plants that are hypersensitive to blue light. This means that irradiation with blue light leads to a stronger inhibition of hypocotyl growth and to a accumulation of anthocyanins. CRY1 therefore fulfils all the criteria for a blue light receptor. It is a soluble cytoplasmic protein which is present in all plant tissues. CRY1 is not the only *Arabidopsis* blue light receptor; CRY2 has been identified as another. This is similar to CRY1, but differs in its physiological effect. Plants that overexpress CRY1 or CRY2 do not show any defects in phototropism or in stomatal closure. Therefore, there must be other blue light receptors. The recently identified NPH1 protein of *Arabidopsis* which is a protein kinase with a putative redox-sensing domain is a

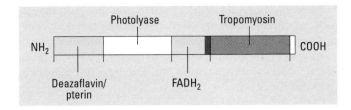

Figure 5.10 Domain structure of the CRY1 (HY4) protein of *A. thaliana*. See text for explanation. (According to Ahmad and Cashmore 1996.)

good candidate for the photoreceptor of the phototrophic response.

Extensive biochemical and genetic characterization of the blue light signal chain is still in its early days. There are specific G proteins in the plasma membrane of the plant cell which could be activated by blue light, and certain plasma membrane proteins are phosphorylated. It is not clear whether CRY1, CRY2, or another blue light receptor is responsible for this. More experiments are needed before a model for the signal transduction chain of blue light can be postulated.

5.3.2 Blue-light-regulated gene expression

Many results suggest the activation of certain genes by blue light, in both intact plants and cell cultures. These include genes for nuclear- and plastid-encoded proteins of the photosynthetic apparatus, enzymes of flavonoid metabolism, and water transport proteins in the plasma membrane. In a similar fashion to the phytochrome-regulated genes, some blue light genes are activated quickly (within 1–2 hours), while others are induced more slowly.

Some of the genes investigated so far are driven not just by blue light but, depending on developmental stage, are also regulated by UV or red light. In etiolated pea seedlings, for example, the *rbcS* genes are activated by the phytochrome system. Blue light is not important at this stage of development. However, if fully grown seedlings are adapted to growth in the dark, the *rbcS* genes can only be activated by blue light. Similar results have been obtained in transgenic petunias which express reporter genes under the control of the pea *rbcS* promoter.

A development-dependent switch from one photoreceptor to another also occurs with the chalcone synthase gene. In *Sinapis alba* the expression of the corresponding promoter in the cotyledons is regulated mainly by phytochromes. In later states of development, the blue light receptor takes over this function.

The search for DNA sequence motifs that are specifically responsible for blue light regulation of promoters has been unsuccessful so far. The 'light switches' that have been identified in the chalcone synthase gene (Figure **5.7**) can be triggered by either red or blue light, so we will have to wait to see whether (and if so, which) genes contain blue-light-specific switches.

5.4 The biological clock

5.4.1 Some basic observations

The pioneering work by the German plant physiologist E. Bünning have made the biological clock (internal clock) and the results of its activity—circadian rhythm—a well-known phenomenon. The term *circadian rhythm* is used to describe certain life processes, such as the movement of leaves, which have rhythmic patterns with a period of 1 day, and which keep their rhythm even if environmental conditions stay constant (free running). This means that an endogenous clock, or oscillator, must exist—a clock that has a period of 24 ± 3 hours (hence the term, circadian). For this rhythm to qualify as circadian, two further conditions must be fulfilled: the period length should be, as far as possible, temperature-independent; and external cues, such as light and temperature, should be able to entrain the rhythm.

An external stimulus is required to start the biological clock, and for plants, the most important external stimulus seems to be light. Switches from light to dark and vice versa are both registered, and the internal clock's periodicity is altered accordingly. This ensures that responses that depend on the biological clock's circadian rhythm occur at specific times of the day. This means that the inner rhythm of the plant can be synchronized to environmental conditions. Because plants register both light on and light off signals (and thus measure day length) they can ensure that certain events, for example flowering, are timed to occur in certain seasons (*see* section 6.3.1).

The biological clock can also be adjusted by diurnal temperature differences. Temperature is the overriding factor for circadian-regulated synthesis of, for example, the polyamine putrescine, or adaptation of the plant to different temperatures. It was observed more than 60 years ago that tomato plants die if they are kept in constant light and temperature conditions. Only by measuring temperature cycles could the plants orient themselves in the day, and synchronize their inner functions with the outside environment.

5.4.2 The biological clock drives gene expression

The plant's physiology is not the only process controlled by the biological clock, so are plant genes, especially light-regulated genes such as those that code for some of the components of the photosynthetic apparatus. The amount of mRNA of these genes fluctuates as the day progresses. One good example of this is the mRNAs for the proteins of the light-harvesting complex of photosystem II (LHC-II). The amounts of these mRNAs can differ by up to 100-fold during the course of the day. Other mRNAs may only fluctuate slightly. Figure **5.11** shows that particular mRNAs have characteristic maxima during the day. The early light-induced-protein (ELIP) mRNA peaks at 8 o'clock, the LHC-II mRNA peaks at 10 o'clock, and the regulatory genes for the water splitting apparatus (OEC) of photosystem II are maximally expressed at about 12 o'clock. Other mRNAs, for instance for a cysteine protease, start to accumulate as night begins. From experiments using light interference, it is known that phytochrome not only starts mRNA oscillation but also is able to shift the phase.

Run-on experiments with isolated nuclei have shown that differences in the amounts of mRNAs are accompanied by oscillating transcription rates. Transcription of these genes is therefore under circadian control. The question is which *cis*-regulatory elements in the promoters of these genes are responsible for the link to the biological clock. The results of some detailed studies of the promoters of LHC-II genes of mono- and dicotyledonous plants suggest that the *cis*-regulatory elements for circadian control and phytochrome control are very near to each another, and may even be identical, or influence each other. This means that the signal transduction chain for the biological clock and light both have the same target in the transcription control machinery.

The widest accepted model for a circadian oscillator requires a negative feedback-loop composed of several state variables (A, B, C, D; Figure **5.12**). The features of these

Figure 5.11 Activities of plant genes during the day. Early in the day the genes for ELIP (*early light-induced proteins*) and for components of PS I and II (LHC II, light-harvesting complex; OEC, regulatory proteins for the water splitting apparatus of 34, 23, and 16 kDa) are transcribed. Other activities occur during the transition to the dark phase or at its end (cysteine protease, thionin, protochlorophyllide oxidoreductase (POR), and glutamyl tRNA reductase (an early gene of chlorophyll biosynthesis)).

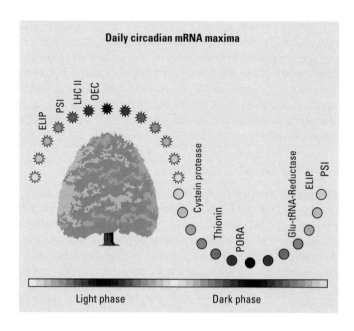

variables are formally described by the parameters a, b, c, and d. Since A oscillates because of negative feedback, the other variables also oscillate in sympathy. From a biochemical standpoint, one could suppose that the oscillator is a protein which negatively autoregulates its own synthesis by transcription, translation, post-translational modification, and transport back into the nucleus (Figure **5.12**). If the individual biochemical parameters, a,b,c,d, run with an appropriate delay, a periodicity of 24 hours is reached.

In the fruit fly *Drosophila melanogaster,* and the fungus *Neurospora crassa*, a number of mutants with defects in circadian rhythm have been isolated and molecularly characterized. *PERIOD (PER)* and *TIMELESS (TIM)* in *Drosophila* and *FREQUENCY (FRQ)* in *Neurospora* are genes which, when mutated, cause errors in circadian rhythm, and can even lead to a complete lack of rhythm. The expression of these genes also oscillates and is negatively autoregulated. *PER, TIM,* and *FRQ* are therefore components of the biological clocks of these organisms. Recently another clock gene of *Neurospora* has been identified—*WHITE COLLAR2 (WC2)*. This is a positive regulator of the expression of *FRQ* and is driven by the light signal chain. *WC2* is therefore a protein that connects the circadian oscillator with light perception.

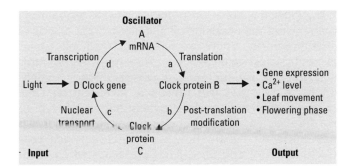

Figure 5.12 A process model of the biological clock composed of the input signal transduction chain, the circadian oscillator, and the output signal chain. As an input signal, only the light is used for the example. It is the most important timer for the adjustment of the biological clock. The input signal is able to interfere at different sites of the oscillatory feedback—in *Neurospora* by the transcription factor WC2. *PER*, *TIM*, and *FRQ* are examples of clock genes. The output signal chain in plants triggers a variety of physiological processes and induces circadian rhythm. (According to Anderson and Kay 1996.)

In plants little is known about the molecular structure of circadian oscillators and their connection to the output signal chain. Mutants that have defects in their circadian rhythm have been described only recently in cyanobacteria and higher plants. Isolation of the affected genes will be the first step towards molecular identification of the components of the biological clock of photosynthetic organisms. These investigations will help our understanding of the biological clock.

5.5 Phytohormones

Auxins were discovered in the 1930s, and there quickly followed a frantic search for similar molecules that act as chemical messengers, directing the physiological activities of plants and acting as a messenger service between the various organs and tissues within the plant. Such messengers seemed to act like animal hormones, and are active at very low dosages (*c.* 10^{-6}–10^{-8} M), therefore they were called *phytohormones*.

Definition Phytohormones are low molecular weight substances produced by plants. They act specifically in low concentrations, usually in a regulatory fashion, across cell borders, and are not chemically modified as they act. They bind to specific receptors, and as they do so, trigger an effect. Phytohormones act as transportable messengers, exchanging information between organs and tissues. In many cases, they also act as locally restricted signal carriers during changing biotic and abiotic environmental conditions.

The following discussion is not intended as a systematic run down of the physiological and biochemical aspects of phytohormones. There are numerous textbooks about plant physiology that already provide this. After an introduction to the various classes of phytohormones (sections 5.5.1 and 5.5.2) we will attempt to shed light on some examples of phytohormone activities that are relevant to the developmental biology of plants: how plant cells recognize hormone signals, how they move between cells (section 5.5.3), and how genes are activated by them (section 5.5.4).

5.5.1 The 'classical' phytohormones and their effects—a brief overview

Auxins, gibberellins, cytokinins, abscisic acid, and ethylene are the five 'classical' groups of phytohormones (Table **5.3**). The first three classes each contain several closely related substances, each of which are effective in laboratory tests, although they may not be as effective in the living plant. There is a large amount of heterogeneity, especially among the gibberellins—more than about 100 different molecules have been assigned to this group. Experiments cannot easily determine whether a given molecule will be effective *in planta* because a molecule from a phytohormone class can be altered by the plant into a different molecule of the same class. A detailed analysis of hormone synthesis mutants will possibly resolve this problem.

Phytohormones do not only exist within the plant freely, they mainly form conjugates. The conjugands can be sugar or amine groups, and are usually connected via an amine or glycosidic bond to the phytohormone. Therefore the concentration of active hormones depends not only on their rate of synthesis and degradation, but also on the equilibrium of free and conjugated forms. A gene for amido-1-hydrolase, which splits indole-3-acetic acid from its conjugate has recently been isolated by positional cloning from *Arabidopsis*. The phytopathogenic bacterium *Agrobacterium rhizogenes* also interferes with the equilibrium of free and conjugated hormone, because it contains a hydrolase-encoding gene in its T-DNA. In this way it changes the concentration of cytokinin.

Phytohormones cause many effects that can overlap between classes. However, phytohormone effects are specific and phytohormones are not interchangeable with

Table 5.3 Classes of phytohormones and their effects.

Phytohormone	Chemical structure	Some effects
Auxins (indole -3-acetic acid)		Stimulation of growth by elongation Formation of adventitious and side roots Maintenance of apical dominance Inhibition of leaf abscision Parthenocarpy
Cytokinins (zeatin)		Stimulation of cell division Lateral-shoot induction Retardation of senescence
Gibberellins (GA3)		Promotion of shoot growth Induction of flower formation in rosette plants Breaking seed dormancy
Abscisic acid		Enhancement of fruit and leaf abscision Inhibition of seed germination Water stress and stomatal closure
Ethylene		Enhancement of fruit ripening Enhancement of fruit and leaf abscision
Octadecanoids (jasmonic asid)		Induction of wound and defence genes Induction of tendril-like behaviour Induction of tuber formation in potatoes Stimulation of secondary metabolism in cell culture
Oligosaccharide (oligogalacturonide)	α-1,4 α-1,4 α-1,4 GalA—GalA—GalA—GalA- - -	Induction of defence reactions by pathogen attack Auxin antagonism Activation of cell division in tobacco protoplasts
Brassinosteroids (brassinolides)		Promotion of cell elongation and division

Peptide hormones
  Systemine
  (tomatoes)
  ENOD40
  (soybeans)
	AVQSKPPSKRDPPKMQTD
	
	MELCWLTTIHGS
		Systemically acting signal of defence against pathogens
		Induction of dedifferentiation of root cortex cells in legumes
		Auxin tolerance in tobacco protoplasts

In cases of auxins, cytokinins, octadecanoids, oligosaccharides, and brassinosteroids a typical example is shown. GalA, galacturonic acid. Single-letter codes are used for peptide sequences.

135

one another. For example, auxin increases the elongation of the graminaceous coleoptile but has no effect on internode lengths. Growth of the latter is controlled by gibberellins, which in turn have no influence on coleoptile growth.

One further characteristic of phytohormones is that they can antagonize each other. The ratio of auxin to cytokinin is important for controlling organ development in tissue culture: an excess of cytokinin leads to the production of shoots by calli; induction of roots is caused by an excess of auxins. Gibberellins and auxins can cause a sex change in some monoecious (*Cucumis sativus*) and some dioecious (*Cannabis sativa*) flowers: application of gibberellins leads to the masculinization of the flower, whereas application of auxin promotes production of female flowers. That this antagonistic effect can also be seen at the level of gene regulation is shown by the antagonistic regulation by abscisic acid and gibberellins of α-amylase activity in the caryopsis of grasses (*see* section 5.5.4).

5.5.2 The new phytohormones: jasmonates, oligosaccharides, brassinosteroids, and peptide hormones

Besides the five classical hormone groups, there are now four more known groups of plant growth regulators: octadecanoids (jasmonates), oligosaccharides, brassinosteroids, and peptide hormones. These have been characterized biochemically, genetically, and physiologically so they can be classified as phytohormones. The spectrum of phytohormones therefore now reaches from low molecular weight substances to oligosaccharides to peptides, so they have a structural variety comparable with the animal hormones.

Octadecanoids

Jasmonic acid has been found in the form of a methyl ester in essential oils of jasmine and rosemary. It is known that jasmonates are ubiquitous throughout the plant kingdom. In higher plants they can be found throughout the whole body of the plant, although they occur mainly in growing tissues. External stimuli such as wounding, mechanical force, plant pathogens, or osmotic stress lead to a higher endogenous jasmonate concentration. The biosynthesis of jasmonates via the octadecanoid pathway starts with linolenic acid, which is initially oxidized by lipoxygenase. Further enzymic steps occur until finally a stable, physiologically active octadecanoid, 12-oxo-phytodienic acid, is formed. The final product of the octadecanoid pathway is (+)-7-iso-jasmonic acid, which is easily isomerized to (−)-jasmonic acid (Figure **5.13**).

Jasmonic acid has many effects (Table **5.3**). Genes that are induced by wounding or pathogenic attack, such as protease inhibitors or polyphenol oxidases, can also be activated by jasmonic acid or other octadecanoids. Mutants that have defects in the octadecanoid pathway are highly sensitive to pest attack. Therefore, jasmonic acid is an important messenger substance for fending against pathogenic attack (Figure **5.13**).

The participation of jasmonic acid or 12-oxo-phytodienic acid in the regulation of the directional growth of a tendril of *Bryonia dioica* is biologically fascinating. If the tendrils of these plants hit a hard surface, such as the shoot of another plant, they bend and a sequence of specific morphological changes occurs. During this time the jasmonic acid levels rise dramatically. Exogenously applied jasmonic acid or 12-oxo-phytodienic acid also lead to this bending reaction. This shows that octadecanoids are signal transducers for touch.

Volatile substances such as jasmonic acid methyl ester are important for communication between different plants, even across species barriers. Indeed, under laboratory conditions, jasmonic acid methyl ester that has been secreted into the air by *Artemisia* leaves, induces the synthesis of proteinase inhibitors in neighbouring tomato plants. Such allelopathic interactions are ecologically very important, and a wide range of these interactions can be investigated using the methods of modern biochemistry and molecular biology.

Oligosaccharides

These signal transducers were first discovered during investigations into plant pathogen defences. When a phytopathogenic fungus attacks a plant the plant produces β-glucanases and chitinases, leading to the release of oligosaccharides and chito-oligosaccharides from the fungal cell wall. These compounds act as elicitors and induce the plant's defence reaction. The octadecanoid pathway acts as the signal transducer (Figure **5.13**). Oligosaccharide elicitors are also formed when polygalacturonidases and pectin lyases from the pathogen act on the pectins within the plant cell wall, hydrolysing them to oligogalacturonides. Oligogalacturonides are also regulators of growth and development. They can hinder root formation when they are applied to explants of tobacco leaves. If the auxin concentration rises, the inhibitory effect of the oligogalac-

Figure 5.13 Jasmonates as signal transducers for the induction of defence genes. See the text for explanation. (According to Bergey *et al.* 1996.)

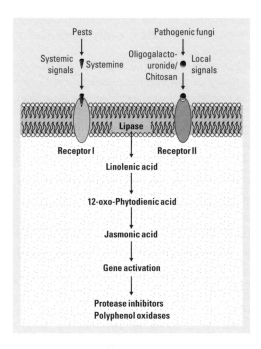

turonide will be overcome, and the explants then form roots. Oligogalacturonides and auxins are thus antagonistic to one another.

The lipochito-oligosaccharides of rhizobia (*see* section 7.3.3) are another good example of the morphogenetic potential of oligosaccharides. However, their biological effect is not limited to the rhizobia–legume symbiosis. Synthetic lipochito-oligosaccharides in femtomolar concentrations can induce division of tobacco protoplasts and replace cytokinins and auxins. Exogenously applied lipochito-oligosaccharides are also capable of complementing the defect in a carrot mutant which is otherwise incapable of somatic embryogenesis. Oligosaccharides in their varied forms are therefore essential components of the signalling network of plants.

Brassinosteroids

These were isolated in the 1970s from pollen extracts of *Brassica napus*. Physiological studies showed that these components can lead to a broad spectrum of effects when they are applied exogenously to whole plants or epicotyl segments. Brassinosteroids stimulate growth by enhancing cell division or cell elongation. They have only recently been classified as phytohormones, after mutations in their biosynthesis pathway (*DET2*: steroid 5α-reductase; *CPD*:

teasterone-23-hydroxylase) in *Arabidopsis* left no doubt that these substances are essential for plant growth and development.

Peptide hormones

These are essential elements in the animal signalling network. With the discovery of two peptides that function as signals, systemine and ENOD40, it was shown that plants also use these molecules to transduce signals.

Systemine

An 18 amino acid long peptide (Table **5.3**), systemine has been identified as a signal substance for the systematic activation of defence genes after tomato plants were attacked by pests. Systemine is proteolytically produced from a larger precursor and is transported by the plant's vascular system. It activates the octadecanoid pathway and therefore the defence genes of the plant. Systemine and oligosaccharide elicitors therefore act by the same signal transducer, the octadecanoid pathway (Figure **5.13**). However, the effect of the oligosaccharide is local, whereas systemine leads to systemic effects.

ENOD40

This was originally identified as a plant gene induced by the *nod* factors of rhizobia (*see* section 7.3.3) in nodule-forming legumes. It is expressed locally in the root pericycle where later the nodule primordia will develop in the cortex. This led to the assumption that its gene product participates in the induction of the nodule primordia, stimulating cell division by influencing the balance between cytokinin and auxin. Transformation experiments with tobacco confirmed this assumption, by showing that ectopically expressed soybean ENOD40 led to the formation of adventitious shoots, reducing apical dominance. ENOD40 transgenes, when controlled by a strong promoter, made it possible for tobacco protoplasts to divide in the presence of a strong auxin concentration. The ENOD40 gene product is therefore an auxin antagonist. The active agent of the ENOD40 gene product is a small, 10–12 amino acid long, peptide which not only occurs in legumes, but also in tobacco. It is the first example of a peptide plant growth regulator.

5.5.3 In search of receptors and signal transduction chains

For a phytohormone to cause a reaction, the cell, using specific receptors, must first recognize the hormone.

Definition Receptors are functionally defined by two criteria: they bind specifically, reversibly, and with high affinity to a phytohormone molecule. On the formation of a phytohormone–receptor complex they can start a signal transduction chain in the cell that will lead to physiological reactions.

The receptors of steroid hormones are inside the cell, so these hormones must cross the cell membrane to reach the receptor. However, most animal hormones, such as insulin and adrenaline, do not enter the cell. Their receptors are on the outer side of the cell membrane. After the hormone binds to the receptor, activating it, a signal transduction chain transmits the signal into the cell (Figure **5.14**). We have a fairly detailed understanding of the receptors and signal transduction chains in animal cells, and the signal transduction paths used by bacteria to recognize environmental conditions, are also well understood. The signal transduction paths of bacteria are two-component systems, each composed of a sensor kinase and a response-regulator protein. For example, the two-component system of VirA and VirG in *Agrobacterium tumefaciens* regulates the activation of the other *vir* genes by phenolic plant substances (*see* section 7.2.2).

To identify phytohormone receptors, two strategies, one biochemical and the other genetic, are employed. The biochemical approach utilizes the phytohormone as a specific 'bait'—this is made possible because the hormone binds with high affinity to its receptor. The phytohormone (or a functional equivalent) is bound as a ligand to the matrix of a chromatography column. Using affinity chromatography, one can search for the protein that binds to the hormone. Another biochemical technique uses photoaffinity labelling to find the protein. A radioactively labelled phytohormone is attached to a photoreactive group that is activated by light to produce a covalent link to the attached protein. Using either of these techniques, one should bear in mind that the proteins found will not necessarily be the receptors, but could also have other functions, such as transporting the hormones.

The biochemical approach has proven especially successful in searching for auxin-binding proteins. Several of these (ZmABP1–ZmABP4) have been biochemically and molecularly characterized in maize, and they are good auxin receptor candidates. These proteins are mainly present in the endoplasmic reticulum, but a small percentage has also been found to be extracellular, on the outer surface of the plasmalemma. It has been shown that these

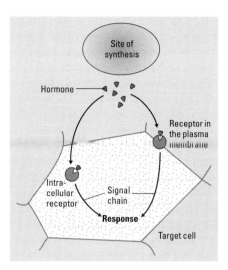

Figure 5.14 Alternative effect chains of hormones.

extracellular proteins are involved in the recognition of the auxin signal. If protoplasts are incubated with antibodies against ZmABP proteins, auxin binding is blocked and the cell's proton-pumping ATPase cannot be activated by auxin so no proteins are pumped out of the cell. Binding proteins have also been described for cytokinins and gibberellins. However, as with the auxin-binding proteins it is not yet known whether these proteins are the actual receptors. There are no certain clues about the signal transduction pathways of these three phytohormones.

A genetic approach has been successful with *Arabidopsis*, where it has been used to characterize the signal transduction pathway of ethylene. If seedlings of *Arabidopsis* are grown in the dark and supplied with ethylene, a characteristic response is observed: growth of the hypocotyl and the root is inhibited, the subapical hypocotyl thickens and the hypocotyl hook bends more than normal. This reaction probably helps the seedlings in their natural environment, to either grow around or push away obstacles they come across in the soil. For scientists, this *triple response* is a very simple system that can be used to look for mutants with defective responses to ethylene. The mutants of *Arabidopsis* that have been characterized fall into two groups. The *ein*, *etr*, and *ain* mutants do not show the triple response in the presence of ethylene, so they must have a defect in the signal transduction pathway from phytohormone to gene expression. The other mutants, *ctr* and *eto*, show the triple response *even in the absence* of ethylene. Therefore, they are either constitutive in hormone production (*eto* mutants) or in their appearance (*ctr* mutants).

Genetic analyses of these mutants has allowed a model to be proposed for the ethylene signal transduction chain (shown in simplified form in Figure **5.15**). In the meantime, four genes, *ETR1*, *ERS1 EIN3*, and *CTR1*, have been cloned. The *ETR1*-encoded proteins share some similarities with prokaryotic sensor kinases. These proteins function in bacteria as both receptors and signal transducers. A second sensor kinase has been identified with *ERS1*, which has 70 per cent sequence similarity to the ETR1 protein. If these genes are functionally expressed in yeast, the plasma membrane of the transgenic yeast binds ethylene with high affinity. Therefore *ETR1* and *ERS1* probably encode ethylene receptors.

CTR1 is a negative regulator of the ethylene response. This means that mutations in this gene lead to constitutive activation of the ethylene signal pathway. The *CTR1*-encoded protein has similarities with the protein kinases of the *Raf* family, which are functionally equivalent to the MEK kinases. MEK and Raf kinases, with the ME and MAP kinases, are a protein kinase cascade which is conserved in multicellular eukaryotes and yeast. It is called the *MAP kinase cascade*. There are several functionally different MAP kinase cascades present in eukaryotes. They are important in the transduction of developmental and stress signals. For example, in yeast they participate in mating type behaviour and osmoregulation. The sequence of the EIN3 protein does not allow a conclusion to be drawn about its function in the signal chain. In any case, it is a nuclear protein which possesses sequence motifs characteristic of transcription factors.

With the sensor kinases ERS1 and ETR1 and the Raf kinase CTR1, the signal transduction chain of ethylene contains two characteristic components of the signal transduction chain of osmoregulation in baker's yeast. The molecular characterization of the other mutants will show whether components similar to the prokaryotic response regulator are following ETR1 or ERS1 and whether the MAPK cascade is a part of the signal transduction chain.

5.5.4 Phytohormones can switch genes on

That phytohormones can turn genes off was first noticed when the regulation of storage metabolism in the caryopsis of grasses was studied. After the caryopsis has germinated, storage substances held in the endosperm and the aleurone layer are mobilized by a hormonal signal emanating from the embryo. This hormone is a gibberellin that induces

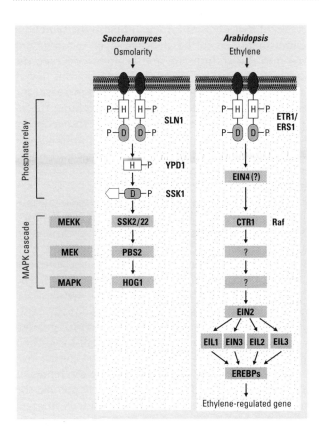

Figure 5.15 The signal transduction chain of ethylene in *Arabidopsis thaliana* in comparison to the regulation of osmolarity in yeasts. An increased osmolarity in the medium in yeast leads to autophosphorylation of a conserved histidine (H) in the histidine kinase domain of the SLN1 protein which belongs to the family of sensor kinases and acts as a dimer. The phosphate group is then transferred to a aspartate (D) in the receiver domain of the SLN1 protein. Finally it is transferred via an intermediary YPD1 protein to the receiver domain of the response regulator, SSK1. This phosphate relay with successive phosphorylations of histidine and aspartate residues often occurs in prokaryotes, for example during the regulation of the sporulation of *Bacillus subtilis* or the control of the virulence factor in *Bordetella pertussis*. SSK1 starts the MAP kinase cascade (MAP: mitogen-activated protein). This is formed by the MAP kinase (MAPK), HOG1, the ME kinase (MEK:

MAP/ERK kinase), PBS2, and the ME kinase kinase SSK2/22. Using further steps, glycerin synthesis is turned on and in this way the osmotic values inside and outside the cell equalize. In *Arabidopsis* the signal transduction chain for the ethylene response starts with the ETR1 and ESR1 proteins. Both are homologous with the SLN1 protein. Where EIN4 stands in this signal transduction chain is unclear. However, it must be above CTR1 which is positioned upstream from EIN2. Gain of function mutations indicate that EIN2 is above EIN3 and EIL1 (*EIN3-like*). As putative transcription factors, EIN3 and the homologous proteins EIL1/2/3 could control other transcription factors (EREBPS: ethylene responsive element binding proteins). Those transcription factors would interact with the known *cis*-regulatory element in the promoters of ethylene-regulated genes. (According to Ecker (1995) [ethylene] and Wurgler-Murphy and Saito (1997) [osmoregulation in yeast].)

production of hydrolytic enzymes, at first in the scutellum epithelium then later in the aleurone layer. α-Amylase is secreted into the endosperm, and begins to degrade the starch stored there. Studies on isolated aleurone layers have shown that incubation with gibberellin increases the amount of α-amylase mRNAs, thereby increasing the α-amylase protein concentration. Run-on transcription experiments have shown that the increase in mRNA levels is due to a stimulation of transcription. Abscisic acid antagonizes the effect of gibberellin, and hinders the activation of gene expression.

The induction of α-amylase gene expression by gibberellins happens only after a delay of several hours. However, auxin-dependent activation of gene expression can be much faster. If pea epicotyl cells are incubated with auxin, cell expansion starts after 15–25 minutes. After just 5–10 minutes, transcription of several genes starts. The increase in transcription is not due to protein synthesis, so the signal for it must be transported along an already existing signal chain. Some of these early auxin-regulated genes code for nuclear proteins with short half-lives. Their amino acid sequences have a β-sheet and two consecutive α-helices, typical of the prokaryotic transcription repressor proteins of the *Arc* type. This would seem to imply that the products of the early auxin-regulated proteins interact with the late auxin-regulated genes which are necessary for the eventual physiological response.

The promoters of the early auxin-regulated genes of pea have two *cis*-regulatory elements that are both necessary for the induction of transcription by auxin and act co-operatively (Figure **5.16**). If the sequence motifs that bestow auxin regulatability are added to a heterologous promoter, this promoter becomes auxin-regulatable. Hormone-responsive *cis*-regulatory elements have also been detected

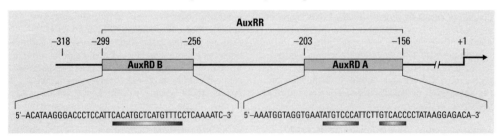

Figure 5.16 Auxin regulator signals in the promoter of the *PS-IAA4/5* gene of pea. The region that is needed for the regulation of the promoter by auxin (AuxRR) is composed of two separate sequence areas (AuxRDA and AuxRDB). The underlying sequence motifs of both elements can also be found in the promoters of other auxin-regulated genes. (According to Ballas and colleagues).

in the promoters of abscisic acid- and gibberellin-regulated genes. However, in all cases studied to date, the *trans*-regulatory factors interacting with these *cis*-regulatory sequences still need to be identified unequivocally. As with light-regulated gene expression, we are still a long way from being able to establish a detailed mechanistic model for transcription activation of hormone-regulated genes.

5.6 From the seed to the young seedling: phytohormones and light as regulators of development

Seeds of cereals and legumes are the staples of the human diet, so seed development takes on a particular importance in plant science. However, to the developmental biologist, study of the development of the seed and the young seedling are well worthwhile. This is because they make ideal subjects for examining the interactions between external factors (such as light and temperature) and internal developmental programmes. Phytohormones, especially abscisic acid and gibberellins, are important in the regulation of these programmes.

Seed maturation and the biogenesis of the photosynthetic apparatus in the young seedling necessitate large-scale protein synthesis. Genes for the required proteins have a high transcription rate, and precisely when and where they are transcribed is very strictly regulated. For this reason they make model genes for study by molecular biologists eager to understand transcriptional regulation of plant genes. Below, we will describe the production of mature seeds through to the development of the young autotrophically growing seedling, paying particular attention to the regulatory mechanisms triggered by light and phytohormones.

5.6.1 In angiosperms, seed development is divided into several phases

Definition The term *seed* describes the mature seed, which includes the embryo, a more or less developed endosperm, and the seed coat (testa).

The development of seeds by angiosperms is initiated (except in the case of the apomixis) by double fertilization. When the egg cell fuses with one of the two sperm cells, the zygote is produced. The zygote then develops into the embryo (*see* section 6.1). The second sperm nucleus fuses with the diploid endosperm nucleus, and after a series of multiple divisions, a multicellular triploid endosperm is produced. The endosperm of many diploid plants is only transient, and is later degraded or absorbed by the embryo. A persistent endosperm, kept for its storage capabilities, is characteristic of the caryopses of grasses.

The processes leading to the production of the dormant seed can be divided into two phases. In the earlier phase, a multicellular embryo with a root, a shoot, and cotyledons develops (see section 6.1). When the cotyledon stage is reached, cell division stops, and from now on, growth of the seed is due to an increase in cell volume and mass. In many angiosperms the embryos of the cotyledon stage possess completely developed chloroplasts.

The later phase of seed development in dicotyledonous plants can be divided into four stages: maturation, post-abscission, pre-desiccation, and desiccation. The stages differ both physiologically and in patterns of gene expression. We will discuss these stages in cotton, in which they have been well studied (Figure 5.17).

During the maturation phase, the embryo reaches its maximum size. Already in the final half of the cotyledon stage, the amount of water in the embryo decreases. This decrease continues until it reaches a plateau in the middle of the maturation phase. The decreasing water potential is accompanied by an increase in abscisic acid. In dicotyledonous plants, the amount of abscisic acid reaches a double-peaked maximum in the first half of the maturation phase. Later, as seed development proceeds, the amount of abscisic acid decreases. In the embryos of grasses, abscisic acid reaches its peak in the post-abscission phase. Genetic investigations with Arabidopsis have shown that the first abscisic acid peak is due to the mother plant; only the second can be attributed to the embryo itself. During the maturation phase, massive production of proteins and lipids is characteristic (Box 5.2). The Gramineae also produce storage carbohydrates, depositing them in the endosperm during this phase.

When the funiculus which connects the ovary with the mother plant is interrupted physically or, at least, functionally, the seed's maturation is complete, and the post-abscission stage begins. In angiosperms with green embryos, the chlorophyll degrades and the integuments brown and harden. The mRNAs for seed storage proteins disappear, and the amounts of LEA (late embryogenesis abundant) mRNAs rise greatly. Because of their different accumulation characteristics, two types of LEA transcripts can be distinguished. The true LEA mRNAs start to accumulate in the post-abscission phase. LEA-A transcripts already start

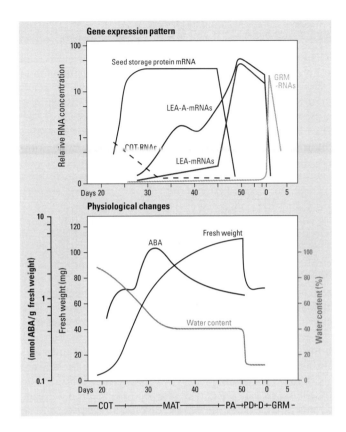

Figure 5.17 The development of seeds in cotton. Seeds ripen in 56 days post-anthesis (DPA). The early phase of embryo and seed development extends over approximately 26 days from anthesis and has been named the cotyledon phase (COT). The late phase of embryo development has been divided into four stages by Galau *et al.*: maturation (MAT), 26–45 DPA; post-abscission (PA), 46–50 DPA; pre-desiccation (PD), 51–53 DPA; and desiccation (D), 54–56 DPA. Cotton has, in common with many other dicotyledonous plants, a transient endosperm and stores the proteins (vicilin, legumin) and lipids in the cotyledons of the embryo. Physiological changes that occur during the ripening of the seeds, such as fresh weight, water content, and level of abscissic acid (ABA), are shown in the lower section of the figure; the upper part illustrates the changes in gene expression. The latter are based on relative concentrations and drawn as the logarithmic values of RNA levels. The simplified results from the abundant RNAs show five time-related patterns of expression. They can be correlated with the cotyledon phase (COT-RNAs), maturation (MAT-RNAs (mainly seed storage proteins) and LEA-A-mRNAs), post-abscission (LEA- and LEA-A-mRNAs), and the germination phase (GRM-RNAs, i.e. the enzymes of the glyoxylic acid cycle). (Adapted from Galau *et al.* 1991.)

accumulating transiently during maturation, and their accumulation parallels that of abscisic acid, hence the 'Λ' in LEA-A. Most of the LEA and LEA-A mRNAs code for hydrophilic proteins. Some of these contain characteristic sequences, such as a tandem repeat of 11 amino acids that can form amphipathic helices. The exact role that the LEA and LEA-A proteins play is not known, but they do help protect the cell against the effects of drying. This has been shown in mutant seeds which do not accumulate LEA proteins, and which cannot tolerate the drying that occurs during seed development.

The desiccation phase is the final phase of seed development. Cotton has a transitionary phase before desiccation starts. This is between when the transcription of the LEA genes stops and the desiccation phase has yet to start. The seed is now genetically inactive, metabolism is reduced to a bare minimum, and the testa is completely differentiated. The embryo now starts a resting phase, during which it can wait for long periods of time without damage, until conditions permit its germination.

147

Box 5.2 Storage proteins of seeds

Storage proteins in the seed are a heterogeneous group of proteins that are produced in large amounts during the seed's maturation phase and then deposited in the embryo or endosperm. Besides the storage proteins, other proteins, such as lectins, proteinase inhibitors, and thionins accumulate. The term 'storage protein' describes only those proteins that are used in the seed as repositories for nitrogen and sulphur. Hence these proteins are rich in amides, and in some cases, sulphur-containing amino acids. They are stored in the vacuole and endoplasmic reticulum, and are visible under the electron microscope as protein bodies. The other proteins in the seed are also cannibalized as nitrogen and sulphur sources, but their primary task is to fight pests and pathogens.

Three groups of seed storage proteins can be distinguished: globulins, albumins, and prolamins. Globulins and certain albumins can be found in all seed plants, but prolamins occur only in the grasses. Albumins and globulins are soluble in water and dilute salt solutions, whereas prolamins are hydrophobic and not water-soluble.

Examples of albumin proteins are the napins of brassicas, and the methionine-rich storage albumins of the brazil nut, *Bertholletia excelsa*. The latter protein is an excellent candidate for genetic manipulation to improve the nutritional value of seeds with proteins low in sulphur, by expressing heterologous sulphur-rich proteins. Globulins are designated 7S (for example, the vicilins) and 11S (for example, the legumins) according to their sedimentation constants. The globulins of legumes have been studied extensively. The prolamins also are categorized in two distinct groups: those of the Triticeae (gliadins, glutelins, and hordeins, amongst others) and those of maize and its close relatives (the zeins).

Seed storage proteins are products of small to large gene families. Exceptionally large families with up to 100 members can be found within the gliadins of wheat and the α-zeins of maize. All genes for seed storage proteins are only expressed in the seeds themselves. Only some of the *cis*- and *trans*-regulatory elements responsible for this are known. The *opaque-2* locus of maize is the first example of a gene whose product regulates the expression of a storage protein. The opaque-2 protein is a transcriptional regulator of the basic leucine zipper type. It specifically recognizes a sequence motif, TCCACGTAGA, that occurs within the genes for the 22 kDa zeins.

Definition The resting stage of a seed is termed *dormancy*. Practically speaking, we talk of seed dormancy when the seed cannot germinate, even when environmental conditions mean that it could.

5.6.2 Searching for global regulators of late seed development

The partitioning of seed development into discrete phases, each with a characteristic pattern of gene expression, leads to the assumption that the seed development programme consists of modules, and that each of the discrete phases has a certain degree of autonomy. This poses two questions: what are the global regulators that control the subprogrammes; and how is initiation of the individual subprogrammes timed?

Unfortunately, we can only partly answer these questions. This is because only one global regulator has been characterized molecularly—the ABI3 (abscisic acid insensitive 3) protein, from *Arabidopsis* (as well as a functionally homologous protein from maize, VP1, or viviparous 1). Both genes code for transcription factors (*see* section 4.1.5).

The *ABI3* gene was originally identified as the genetic locus which, when mutated, would cause insensitivity to abscisic acid. Thus, germination of *abi3* seeds is not hindered by incubating them in abscisic acid (*see* section 5.6.3). Meanwhile, several mutated alleles of this gene, differing in the strength of their defects, are known. Seeds with stronger *abi3* alleles never become dormant, and germinate on their mother plant (viviparously). When *abi3* seeds dry, they lose their ability to germinate. A functional ABI3 protein is therefore required for drought resistance.

Certain gene expression programmes are also altered in *abi3* mutants. There is a drastic reduction in the amounts of mRNAs for the seed storage proteins napin (*At2 S3*) and cruciferin (*CRC*), both of which belong to the *MAT* genes. This also holds true for the *LEA* gene *AtEm6* and the *LEA-A* gene *AtEm1*. However, not all the *MAT* and *LEA/ LEA-A* genes in the *abi3* mutants have altered expression patterns.

ABI3 is only expressed in the seed, and then only during the late phase of the seed's development. This explains why *abi3* mutants show only a seed phenotype. If *ABI3* is expressed ectopically in *Arabidopsis* leaves, some napin, cruciferin, or *AtEm1* mRNAs can be detected, which would never be the case for the leaves of wild-type plants. This provides further proof that expression of these genes in the seed is controlled by *ABI3*. If plants that express *ABI3* ectopically are treated with abscisic acid, the amounts of cruciferin, napin, and *AtEm1* mRNAs in the leaves increase dramatically. Wild-type plants treated in a similar fashion show no change in levels of transcription. The expression levels of these *MAT* and *LEA-A* genes are therefore also controlled by abscisic acid, but only if the ABI3 protein is present.

In *ABI3* and its homologue *VP1* in maize, we have regulators that act throughout the late phase of seed development, and that directly control the expression of some (but not all) *MAT* and *LEA/ LEA-A* genes. It is important to bear in mind that *ABI3* and *VP1* are regulated by the signal transduction chain of abscisic acid.

FUS3 is another regulator of seed development in *Arabidopsis*, but its molecular nature is not known. *fus3* mutants belong to the class of *fusca* mutants (*see* section 5.2.7). Dry *fus3* seeds are wrinkled and do not germinate.

However, if the seeds are collected before they dry, they can germinate and become normal plants. As with *abi3* seeds, *fus3* seeds do not acquire a tolerance to drying during their development. However, in contrast to *abi3* mutants, germination of *fus3* mutants can be hindered by abscisic acid. Also, the *fus3* phenotype is limited to the seed only.

Seed storage proteins and their mRNAs cannot be detected in *fus3* mutants, although other abundant seed proteins are not affected by the *fus3* mutation. Transformation experiments with different seed gene promoters have shown that the lack of a functional FUS3 protein specifically hinders the transcription of genes for seed storage proteins. Just like ABI3, FUS3 might be a transcription regulation protein.

With *fus3* mutants, the late phase of seed development is affected, but so, too, is the early phase. Cotyledons of *fus3* embryos already have leaf-like structures such as trichomes, and are shaped like leaf primordia. These features, together with vivipary and defects in the synthesis of seed storage proteins, are common in *fus3* embryos and *lec* (leafy cotyledon) mutants. Investigations of epistatic interactions have led us to conclude that *FUS3* and *LEC1/2* genes are part of the same effector chain, which is distinct from the *ABI3* chain (Figure **5.18**).

A total of four genes, *abi3*, *fus3*, and *lec1/2*, have been identified which act specifically in the seed, having regulatory functions there. Already, we can see that our previous assumption, that for each development phase there is a global regulator, is probably wrong. Moreover, the networks of genetic regulation are far more complex than was thought originally, and further genetic investigations must be conducted before the first valid models for the late phase of seed development can be constructed.

5.6.3 The role of abscisic acid in seed development

A regulatory model of late seed development must take into account the role of abscisic acid, as well as those of the genetic components. It is a matter of controversy whether abscisic acid is a master regulator in the endogenous development programme, or whether it signals the decreasing water potential during seed maturation. Experiments with isolated embryos and with abscisic acid-deficient mutants allow us to make some preliminary conclusions.

If embryos of dicotyledonous plants are isolated from the developing seed when they are in the late cotyledon or early maturation stages, and are then incubated in a basal

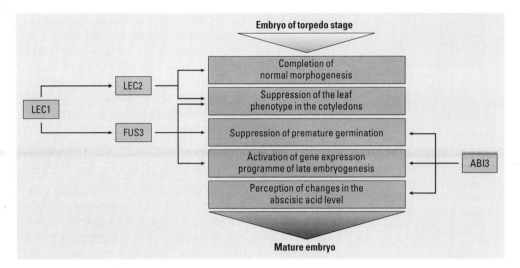

Figure 5.18 A preliminary model of the genetic control of the late seed and embryogenic development of *Arabidopsis*. The key processes of normal seed development are shown in the red boxes. They are not listed in chronological order since they partially overlap or occur in parallel. The arrows show which regulatory genes are required for specific processes. (According to Meinke, D. W., Franzmann, L. H., Nickle, T. C., and Yeung, E. C. (1994). *Leafy cotyledon mutants of arabidopsis. The Plant Cell,* **6**, 1049–64.)

medium, they germinate earlier than normal and go on to develop into normal plants. Therefore, the later phases of embryo development do not affect the ability of the embryo to germinate. Embryos at the late cotyledon stage are already able to germinate. Premature germination is prevented if the media contains abscisic acid, or if the embryo has been water stressed. Thus abscisic acid or a lowered water potential can induce dormancy in the isolated embryo. If dormancy is induced by abscisic acid, it cannot be reversed by gibberellin.

That abscisic acid is also essential for dormancy *in planta* has been shown by experiments with *Arabidopsis* mutants that lack abscisic acid. It was found that the abscisic acid produced by the embryo, but not that produced by the mother plant, induces dormancy.

Can it be that abscisic acid is also the signal that controls the gene expression programme of the maturation and post-abscission phases? If cotton embryos at either the late cotyledon or early maturation stages are cultured on minimal media, they immediately stop transcribing genes for storage proteins and start synthesizing *LEA* and *LEA-A* mRNAs. If abscisic acid is supplemented, the amounts of *LEA* and *LEA-A* mRNAs increase more than they would in the absence of this hormone. This implies that abscisic acid is not necessary for the induction of *LEA* and *LEA-A* gene transcription. The decisive inducer for turning off seed storage protein production and simultaneously turning on the *LEA* genes is the removal of the embryo from its seed. That abscisic acid is not needed for induction of the *LEA* and *LEA-A* genes, but *is* needed for maximum expression of these genes has also been shown during investigations of abscisic acid-deficient *aba1* mutants of *Arabidopsis*.

Abscisic acid therefore has two roles in the late phase of seed development: it is essential for inducing seed dormancy, and it is a quantitative regulator of the *LEA* and *LEA-A* genes. Abscisic acid is the regulator that modulates developmentally driven expression of the *LEA* and *LEA-A* genes. It is not yet known whether the component of gene expression that is dependent upon abscisic acid is regulated ultimately by water potential, or whether other factors are responsible for this.

5.6.4 Germination

Definition The emergence of the radicle from the micropyle is often used to define exactly when germination begins.

Most seeds of wild plants are unable to germinate immediately after they have been released from their mother plant — in other words, they are *dormant*. The embryo can only germinate when blocking mechanisms (which can be either mechanical, physiological, or biochemical) have been removed. However, the seed will only germinate if certain environmental conditions are met. Such criteria are: availability of water and oxygen, and the right temperature and light conditions (which vary from species to species).

Embryo dormancy is usually induced in the late phase of seed development by abscisic acid. However, for continued dormancy several other mechanisms are involved. To break dormancy, environmental factors are often required. Two strategies will be discussed as examples.

Since germination is impossible unless the seed is able to take up water and oxygen, an impermeable barrier around the embryo is an effective way to prevent seed germination. Barriers such as these can reside in the endosperm, nucellus, testa, or the fruit wall. When the seed is stored in the ground, barriers such as these are rotted by soil microbes, so the seed can germinate. Instead, or in addition, to mechanical barriers, chemical barriers to germination can also be used. Any part of the seed can contain chemical inhibitors. One example is amygdalin, a glycoside of hydrogen cyanide. When hydrolysed, hydrogen cyanide accumulates in the embryo and blocks the seed's metabolism. In Rosaceae, only when the endosperm surrounding the embryo rots can the cyanide be released and the germination barrier removed.

Physiological germination barriers are usually broken down by environmental factors. Often, seeds must be kept in the cold for several days or weeks (a process called *vernalization*). The particular physiological processes induced by this cold treatment tend to differ from seed to seed, and none is known at the molecular level.

For many seeds, light is the factor that can break dormancy. Plants that depend on light for germination are called *light germinators*, as opposed to *dark germinators*. Light germinators often tend to have small seeds with small stores of food; seeds that store a lot of food (such as those of most crop plants) are usually able to germinate in the dark. Light and dark germination are usually quantitative, rather than qualitative characteristics. Just how strong is the demand for light during germination depends on the amount of light during formation of seeds on the mother plant. If *Arabidopsis* grows while supplemented with red light as it produces seeds, more of its seeds will germinate in the dark than would the seeds of plants grown in far-red-enriched light. A high P_{fr}/P_r ratio in seeds can abolish the light-dependent nature of germination.

In the presence of gibberellin, the seeds of light germinators are also able to germinate in the dark. It is not yet known whether the effect of light is transmitted via gibberellins, or whether the light and gibberellin pathways are separate but synergistic. That gibberellins are essential to initiate germination has been shown by genetic analyses: mutant seeds with defective gibberellin production are incapable of germinating.

The ecological advantage that seed dormancy confers is clear. Dormancy allows plants, in the form of seeds, to overcome long periods of unfavourable conditions with no ill effects. It is also essential for the plant's success that seed germination coincides with the most favourable conditions for the plant's growth. It is therefore not surprising that plants have developed a broad variety of strategies for seed development and germination.

5.6.5 Early seedling development: from heterotrophy to autotrophy

Since the seed's resources are limited, it is vital for the emerging seedling to become autonomous as soon as possible. Two main events in the early germination phase help the seedling to achieve this aim: storage substances are mobilized in the cotyledons and the photosynthetic apparatus is assembled.

The storage substances of the seed—storage proteins, lipids, and carbohydrates—are the building blocks and energy sources for germination. Their mobilization (that is, their degradation into smaller, usable forms) happens at the sites where they are stored within the cell. Storage proteins in the protein bodies are broken down by endo- and exopeptidases into their constituent amino acids. These are preferentially reserved for *de novo* synthesis of proteins, but can also be utilized in metabolic pathways via the TCA cycle. Storage lipids (mainly triglycerides) are degraded into fatty acids and glycerol in the glyoxysomes. By β-oxidation, the fatty acids are broken down into acetyl-CoA and NADH. NADH is used in the respiratory chain, and acetyl-CoA can be transformed into hexose sugars via the glyoxylate pathway. Finally, storage carbo-hydrates are hydrolysed and phosphorolytically broken down to hexose sugars within the plastids. Most of the enzymes that are required for mobilizing the various stor-age substances are synthesized *de novo* during the seed's germination. In general, genes for these enzymes must be transcriptionally activated (*GRM* genes; Figure **5.17**).

With a slight delay, the seedling starts to develop its pho-tosynthetic apparatus. Since all photosynthetic reactions, with the exception of sucrose synthesis, take place only in the chloroplasts, the development of the photosynthetic apparatus is virtually identical with the development of the chloroplast. From a physiological standpoint, photosynthe-sis in angiosperms is also a feature of a particular organ, the leaf. For this reason, in angiosperms, chloroplast differen-tiation is an integral component of leaf development (Figure **5.19**).

The biogenesis of chloroplasts can be divided into three phases, which are easy to study experimentally in grasses (*see* Box **4.2** and Figure **5.19**). The first steps are activation of plastid DNA replication and multiplication of plastids in the meristematic zone of the leaf (phase I, Figure **5.19**). Only in this zone of the leaf do the cells divide and are imprinted to perform future functions. The activity of the plastid- and nuclear-encoded genes for photosynthesis is still low at this point, but some photochemical activity can already be detected.

Phase II (Figure **5.19**) is the build-up phase. The numbers of chloroplasts per cell increase, the cells no longer divide, and further growth is only through cell elongation. The transcription rates of genes for photosynthesis increase drastically, and the synthesis of the thylakoid membrane complexes and the enzymes of the Calvin cycle increase in parallel. The four large protein complexes of the thylakoid membranes (the reaction centres of photosystems I and II, the cytochrome-b_6/f complex, and ATP synthase) are nearly all synthesized at the same time. The peripheral antenna complexes of both photosystems are added later on. The capacity to fix CO_2 develops more slowly. In the C_4 grass, maize, it has been found that the delayed establishment of light-harvesting capacity correlates with the accumulation kinetics of the carboxylating enzymes Rubisco and phosphoenolpyruvate carboxylase. Similar principles hold true for the C_3 plant, barley. Under normal growth (that is, a diurnal light/dark cycle), the seedling starts by establishing its photochemical capability. The efficiency of photosynthesis is then improved by adding more and more antenna complexes and by increasing the CO_2 assimilation capacity.

A completely different picture emerges if the seedling is first grown in the dark (etiolation) and only later is exposed to the light (greening). Under these circumstances,

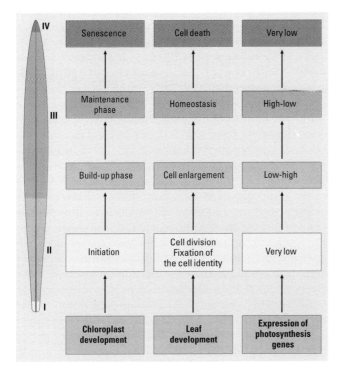

Figure 5.19 A simple process model of the biogenesis of the photosynthesis apparatus in the development of a young grass leaf. The four developmental phases, especially the production and maintenance phase, cannot be clearly distinguished from one another because they form a continuum. Also, the statements about the amount of gene expression are generalized and not necessarily true for every individual photosynthesis gene. See also Box **4.2** and Figure **4.9** for further information about the ultrastructural changes during chloroplast biogenesis. (According to Chory, J. (1991). Light signals in leaf and chloroplast development: photoreceptors and downstream responses in search of a transduction path-way. *New Biology*, **3**, 538–48; and Mullet, J. E. (1993). Dynamic regulation of Chloroplast transcription. *Plant Phasiology*, **103**, 309–13.)

155

synthesis of the thylakoid membrane complexes becomes discontinuous. The cytochrome-b_6/f complex and ATP synthase are already made in the etiolated seedling. However, photosystems I and II will only appear after illumination. The regulatory processes involved in etiolation and greening are therefore not necessarily applicable to chloroplast biogenesis under normal day/night conditions.

Despite these problems, the study of light-induced transition of etioplasts into chloroplasts has led to some valuable insights into the principles of regulation during the biogenesis of multimeric protein complexes. In general, all subunits of these multimeric protein complexes are synthesized together in a co-ordinated fashion. After translation (on either cytosolic or plastid ribosomes) the subunits of these complexes are immediately assembled into functional units. Free subunits are unstable, and so degrade. This holds true especially for the apoproteins of the chlorophyll-*a* protein complexes of both reaction centres. Reconstitution experiments with isolated etioplasts have shown that plastome-encoded chlorophyll apoproteins are only stable in the presence of their ligand, chlorophyll. In fact, chlorophyll is already needed *during* the translation of the apoproteins. It is not yet known whether it has an effect during translation initiation or peptide elongation. Therefore chlorophyll itself is an important regulatory factor for control of translation of plastid genes (*see* section 4.2.8). Since chlorophyll synthesis in angiosperms is strictly dependent on light, the lack of chlorophyll in dark-grown seedlings explains why photosystem I and II complexes are not assembled under conditions of etiolation.

The transition from the build-up phase to the stationary phase (phase III, Figure **5.19**) is a continuous one. Once the cells have reached their maximum size, the photosynthetic apparatus has been built up and the cells become net producers of photoassimilates. The amount of mRNAs from nuclear- and plastome-encoded photosynthesis proteins declines, with the exception of two mRNAs: those for the D1 and D2 proteins of photosystem II remain high. D1 and D2 proteins complex the P680 reaction centre of chlorophyll and the primary and secondary electron receptors. If more electrons are fed into the electron transport chain via photosystem II than can be used by photosystem I for NADPH production, the plastoquinone pool becomes over-reduced. This can lead to damage to the D1 protein caused by radicals (photoinhibition). For this reason the D1 protein, and probably some D2 as well, must be constantly replaced by repair synthesis.

Once its photosynthetic apparatus is complete, the young seedling can grow photoautotrophically. At this stage, early seedling development ends. More importantly though, cells are now able to synthesize new material for growth. The photosynthetic machinery can adapt to work in different light intensities by altering the amounts and compositions of the light-harvesting complexes. Also, protective mechanisms can be invoked against the dangers of light stress. Only in the terminal phase of leaf development (phase IV, Figure **5.19**) is additional synthesis no longer possible. At this point, the mesophyll cells of the leaves and their chloroplasts have started to senesce.

5.7 Bibliography

Phytochrome

Barnes, S. A., McGrath, R. B., and Chua, N. H. (1997). Light signal transduction in plants. *Trends in Cell Biology*, **7**, 21–6.
Summarizes the current state of knowledge in this field

Batschauer, A., Gilmartin, P. M., Nagy, F., and Schäfer, E. (1994). The molecular biology of photoregulated genes. In *Photomorphogenesis in plants*, (ed. R. E. Kendrick and G. H. M. Kronenberg), pp. 559–99. Kluwer, Dordrecht.
Uses rbcS, cab, and chs genes as examples for photoregulated genes.

Furuya, M. and Schäfer, E. (1996). Photoperception and signalling of induction reactions by different phytochromes. *Trends in Plant Science*, **1**, 301–7.
Good addition to Smith (1995).

Mathews, S. and Sharrock, R. A. (1997). Phytochrome gene diversity. *Plant Cell Environment*, **20**, 666–71.
Overview of the structure and evolution of the phytochrome gene family. Good addition to Quail (1994 and 1997).

Quail, P. (1994). Phytochrome genes and their expression. In *Photomorphogenesis in plants*, (ed. R. E. Kendrick and G. H. M. Kronenberg), pp. 71–104. Kluwer, Dordrecht.

Quail, P. H. (1997). An emerging molecular map of the phytochromes. *Plant Cell Environment*, **20**, 657–65.
Detailed reports about the phytochrome proteins and their genes.

Smith, H. (1995). Physiological and ecological function within the phytochrome family. *Annual Review of Plant Physiology and Plant Molecular Biology*, **46**, 289–315.
Highly recommendable review with a lot of depth.

Wei, N. and Deng, X.-W. (1996). The role of *COP/DET/FUS* genes in light control of *Arabidopsis* seedling development. *Plant Physiology*, **112**, 871–8.
Summarizes the current knowledge about the functions of the COP/DET/FUS proteins.

Blue light receptor

Ahmad, M. and Cashmore, A. R. (1996). Seeing blue: the discovery of cryptochrome. *Plant Molecular Biology*, **30**, 851–61.
Good addition to Short and Briggs (1994).

Short, T. W. and Briggs, W. R. (1994). The transduction of blue light signals in higher plants. *Annual Review of Plant Physiology and Plant Molecular Biology*, **45**, 143–71.
Comprehensive review of the signal transduction chains of blue light and blue light-regulated gene expression.

Biological clock

Anderson, S. L. and Kay, S. A. (1996). Illuminating the mechanisms of the circadian clock in plants. *Trends in Plant Science*, **1**, 51–7.
Reviews the current data about circadian rhythms in plants.

Iwasaki, K. and Thomas, J. H. (1997). Genetics in rhythm. *Trends in Genetics*, **13**, 111–15.
Good overview of the circadian clock Drosophila and Neurospora.

Phytohormones

Abel, S. and Theologis, A. (1996). Early genes and auxin action. *Plant Physiology*, **111**, 9–17.
Reviews the molecular natue of early auxin-induced genes.

Appleby, J. L., Parkinson, J. S., and Bourret, R. B. (1996). Signal transduction via the multi-step phosphorelay: not necessarily a road less traveled. *Cell*, **86**, 845–8.
Short review of the organization of multi-step phosphorelays in pro-and eukaryotes

Bergey, D. R., Howe, G. A., and Ryan, C. A. (1996). Polypeptide signaling for plant defensive genes exhibits analogies to defense signaling in animals. *Proceedings of the National Academy of Sciences, USA*, **93**, 12053–8.
The central role of systemine for the systemic activation of plant defense genes is discussed.

Creelman, R. A. and Mullet, J. A. (1997). Biosynthesis and action of jasmonates in plants. *Annual Review of Plant Physiology and Plant Molecular Biology*, **48**, 355–81.
Summarizes our current knowledge of the biology of jasmonates.

Ecker, J. R. (1995). The ethylene signal transduction pathway in plants. *Science*, **268**, 667–75.
The 'classical' review of the signal transduction chain of ethylene.

Wurgler-Murphy, S. M. and Saito, H. (1997). Two-component signal transducers and MAPK cascades. *Trends in Biochemcial Sciences*, **22**, 171–6.
Up to date review of the coupling of two-component systems and MAPK kinases cascades in eukaryotes.

John, M., Röhrig, H., Schmidt, J., Walden, R., and Schell, J. (1997). Cell signalling by oligosaccharides. *Trends in Plant Science*, **2**, 111–15.
Discusses the spectrum of oligosaccharides found in plants.

Kende, H. and Zeevaart, J. A. D. (1997). The five 'classical' plant hormones. *The Plant Cell*, **9**, 1197–210.

Summarizes the knowledge of the bisoynthesis and functions of the 'classical' plant hormones.

Merlot, S. and Giraudat, J. (1997). Genetic analysis of abscisic acid signal transduction. *Plant Physiology*, **114**, 751–7.

Swain, S. M. and Olszewski, N. E. (1996). Genetic analysis of gibberellin signal transduction. *Plant Physiology*, **112**, 11–17.
Short reviews of the signal transduction chains of abscisic acid and gibberellins.

van de Sande, K. et al. (1996). Modification of phytohormone response by a peptide encoded by ENOD40 of legumes and a non-legume. *Science*, **273**, 370–3.
Original work on ENOD40.

Yokota, T. (1997). The structure, biosynthesis and function of brassinosteroids. *Trends in Plant Science*, **2**, 137–43.
Reviews the biosyntheses of brassina steroids and the genetic analysis of their functions.

Seed production, germination, and seedling development

Bewley, J. D. (1997). Seed germination and dormancy. *The Plant Cell*, **9**, 1055–66.
Discusses the physiological basis of seed dormancy and germination.

Galau, G. A., Jakobsen, K. S., and Hughes, D. W. (1991). The controls of late dicot embryogenesis and early germination. *Physiologia Plantarum*, **81**, 280–8.
Uses cotton as an example for studying the late phase of seed development and the role of abscisic acid.

Leech, R. M. (1984). Chloroplast development in angiosperms: current knowledge and future prospects. In *Chloroplast biogenesis*, (ed. N. R. Baker and J. Barber), pp. 1–21. Elsevier, Amsterdam.
Still useful review article of the morphological and physiological changes during the biogenesis of chloroplasts.

Lopes, M. A. and Larkins, B. A. (1993). Endosperm origin, development, and function. *The Plant Cell*, **5**, 1383–99.
Detailed overview of the endosperm.

McCarty, D. R. (1995). Genetic control and integration of maturation and germination pathways in seed germination. *Annual Review of Plant Physiology and Plant Molecular Biology*, **46**, 71–93.
Summarizes the roles of abscisic acid and of Vp1/AB13 for the regulation of seed development.

Ougham, H. J. and Francis, D. (1992). The molecular basis of mesophyll cell development. In *Crop photosynthesis: spatial and temporal determinants*, (ed. N. R. Baker and H. Thomas), pp. 313–36. Elsevier, Amsterdam.
Addition to Leech (1984).

Shewry, P. R., Napier, J. A., and Tatham, A. S. (1993). Seed storage proteins: structures and biosynthesis. *The Plant Cell*, **5**, 945–56.
Focuses on the structure and biosynthesis of seed storage proteins.

Thomas, T. L. (1993). Gene expression during plant embryogenesis and germination: an overview. *The Plant Cell*, **5**, 1401–10.

Our knowledge of *cis-* and *trans-*regulatory elements is summarized.

Weber, H., Borisjuk, L., and Wobus, U. (1997). Sugar import and metabolism during seed development. *Trends in Plant Science*, **2**, 169–74.

The metabolic aspects of seed development are described.

6 Phases during the life cycle of the flowering plant

Contents

6.1 Embryogenesis

6.1.1 The emergence of the basic body plan: pattern formation during early embryogenesis

6.1.2 The growth and differentiation phase

6.1.3 The origin of the primary root meristem and the embryonic root

6.1.4 The origin of the primary shoot meristem

6.2 Postembryonic vegetative development

6.2.1 Development of the primary root

6.2.2 Development of an epidermal cell type in the root: the root hair

6.2.3 Lateral and adventitious roots

6.2.4 Shoot development

6.2.5 The shoot meristem is organized in layers and zones

6.2.6 Leaf development

6.2.7 Development of a cell type in the leaf epidermis: the trichome

6.3 The generative phase

6.3.1 Temperature and photoperiod as inducers of flower formation

6.3.2 The production of a flower organ—an interaction between homeotic cadastral, and meristem identity genes

6.3.3 Production and development of the male and female gametophytes

6.3.4 Pollination, fertilization, and formation of the zygote

6.4 Bibliography

Preface

The life cycle of a flowering plant is divided into three major parts, which are defined by physiological turning points. These parts are embryogenesis, postembryonic vegetative development, and generative development. The embryo develops within the ovule. After this, the embryo and the seed both ripen and begin a period of dormancy, during which the dry seed with the completely developed embryo can overcome unfavourable conditions and, at the same time, can be dispersed. Upon germination the embryo becomes a seedling with primary meristems at the poles (the far ends) of its body axis. These generate the shoot and the root. The shoot meristem mainly produces leaves during the vegetative phase of development. Physiological changes (the induction of flowering) initiate the generative phase. The shoot meristem now produces flowers in which female and male organs make haploid spores by meiosis. The spores develop into gametophytes containing the gametes. This kind of life cycle is the simplest form and occurs in annual plants. In plants that live for several years (perennials) the transition from the vegetative to the generative phase happens repeatedly in each new vegetation cycle.

The organization of the plant's cells influences its development. The cell wall ensures that cells cannot leave their neighbourhood, therefore morphogenesis during development of a plant is due to localized activities of plant cells. On the other hand, the co-ordination of developmental processes relies on long-distance signals (for example from phytohormones). Plasmodesmata can play another part in plant-specific communication. The plant cells form the *symplast* via these plasmodesmatal connections. However, it is not known exactly when during development these cytoplasmic connections are made and which molecules are exchanged thereby.

Different events contribute to development: growth by cell division, production of different cell types by cell differentiation, generation of a body organization by pattern formation, and morphogenesis. Pattern formation describes the process resulting in the functional arrangement of different organs and tissues in space. Morphogenesis describes all cellular processes that are required to give shape to the plant, such as cell division, differential rates of mitosis in different locations within the plant, and directed cell elongation. These different events during development result in a fertile adult plant emerging from a single fertilized egg cell.

In the following section, *Arabidopsis* will be the main model used to analyse the life cycle of a flowering plant. Beginning with the zygote, we will progress through the vegetative adult plant to the production of flowers and gametes. Many, although not all, of the features we will describe for *Arabidopsis* are valid for most dicotyledonous plants, whereas monocotyledonous plants have some special characteristics of their own.

6.1 Embryogenesis

The embryo develops in the ovule of the mother plant, which is connected via a stalk (the *funiculus*) to the wall of the ovary. At fertilization the ovule contains the female gametophyte—the embryo sac that is made up of seven cells (section 6.3.3). Two of these cells, the haploid egg cell and the diploid central cell, will be fertilized. In this way a diploid zygote and the triploid cell (which will later form the endosperm) emerge. The zygote also produces an extra-embryonic suspensor that anchors the young embryo within the ovule and may supply it with substances from the mother plant. In later stages of development the suspensor loses its importance, becoming only a vestigial appendage. Since the suspensor arises from the zygote, it has itself the potential to become an embryo, but the developing embryo appears to send out signals to hinder such development.

Embryogenesis in all flowering plants is more or less similar, even though the seedlings of monocotyledonous and dicotyledonous plants differ in their outer appearance and construction. The major difference between the two is that in dicotyledonous plants two cotyledons flank the shoot primordium symmetrically, whereas in monocotyledonous plants only one cotyledon develops, which will overgrow the shoot meristem, pushing it to one side. We will now describe the embryogenesis of dicotyledonous plants using *Arabidopsis* as an example.

In *Arabidopsis*, starting from the point of fertilization, the development of a mature embryo takes 9 days at a temperature of 25 °C. The following maturation process will need another couple of days. In this time the embryo grows into approximately 20 000 cells, of diameter 500 μm. The zygote is only 20 μm in diameter. The embryo changes shape during growth. Starting as a sphere, it later becomes heart-shaped, then torpedo-like, and finally U-shaped. Accordingly, one can distinguish the different stages of embryo development (globular, heart, and torpedo stages). The early stages are named after the numbers of cells in the pro-embryo (quadrant, octant). Using morphological criteria, the embryo development of *Arabidopsis* is divided into 20 discrete phases (Figure **6.1**). This is possible because cell division in *Arabidopsis* embryos is very regular and the mature embryo has relatively few cells.

After fertilization, the zygote elongates to three times its original length along its future apical–basal axis. After about 10 hours it divides asymmetrically. The small apical cell will eventually give rise to nearly the entire embryo, while the daughter cells of the large basal cell will generate the root meristem and the extra-embryonic suspensor. The apical cell divides three times very regularly. The first two times it divides vertically then horizontally, leading to a pro-embryo of eight cells (the *octant*). The cells of the octant are arranged in two tiers. The upper tier will later produce mainly the cotyledons and shoot meristem, and the lower tier will go on to produce mainly the hypocotyl and the root. The basal cell repeatedly divides horizontally and a file of 6–9 cells emerges. During the octant stage, the cell next to the pro-embryo forms the hypophysis and participates in the development of the embryo. The other cells form the suspensor. This marks the end of the very early phase of embryogenesis.

After the octant stage the embryo grows by cell division. Along the longitudinal axis of the embryo, regional differences become visible. The cells of the upper tier divide with no apparent orientation. A relatively small apical region emerges. However, the cells of the lower tier produce cell files along the longitudinal axis by dividing in a particular orientation. In this way the central region of the embryo enlarges in comparison to the apical region. The basally adjacent hypophysis is the cell from which the basal region of the embryo originates. It emerges by a stereotypic programme of cell division. The hypophysis divides asymmetrically to form an upper lens-shaped and a lower trapezoidally shaped cell. When the heart stage has been reached, the smaller lens-shaped cell has produced four cells by two vertical divisions. Thereafter, these remain mitotically inactive and become the quiescent centre of the root meristem. The larger trapezoidally shaped cell (which abuts the suspensor) generates the initials of the central root cap by two vertical cell divisions. The embryo remains globular during these divisions. Only when the two cotyledon primordia start to grow does the embryo attain a bilaterally symmetrical heart shape.

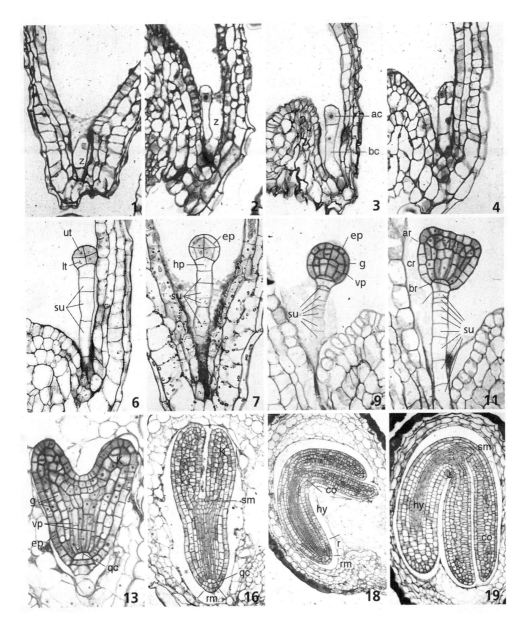

Figure 6.1 Embryogenesis of *Arabidopsis*. Stages of development: (1) zygote; (2) elongated zygote; (3) one-cell stage; (4) two-cell stage; (6) octant; (7) dermatogen; (9) mid-globular stage; (11) triangular stage; (13) mid-heart stage; (16) mid-torpedo stage; (18) bent-cotyledon stage; (19) mature embryo. ac, bc, apical and basal daughter cell of the zygote; ar, cr, br, apical, central and basal region; ep, epidermal primordium; g, ground tissue; hy, hypocotyl; hp, hypophysis; co, cotyledon; vp, vascular primordium; ut, lt, upper and lower tiers of the octant pro-embryo; qc, quiescent centre of the root meristem (rm); sm, shoot meristem; su, suspensor cell; r, root; z, zygote. (According to Jürgens and Mayer 1994.)

Between the octant and heart stages, most tissue layers of the embryo emerge by several periclinal (radial) cell divisions. In periclinal cell divisions, the newly formed cell wall is oriented in parallel to the surface of the embryo, giving an outer and an inner daughter cell. These cell divisions therefore produce new layers successively from the outside to the inside. By periclinal cell divisions in the octant pro-embryo, daughter cells emerge on the outside of the embryo. These constitute the primordium of the epidermis, which is called the *protoderm*. After this the protoderm cells will only divide anticlinally. In anticlinal cell divisions, the newly formed cell wall is oriented at right angles to the surface of the embryo. In this way the outer layer remains intact while the embryo grows. The generation of further tissue layers is now limited to the central region of the embryo. The inner mass of cells is dividing in the early globular stage: the cells in the centre are precursors of the vascular system (procambium), and the cells that surround them will form the ground tissue. In this way within a short period the main tissues are established as concentric layers which are perpendicular to the embryonic axis. The ground tissue will now divide anticlinally. However, the procambium will divide periclinally, producing the precursors of the xylem and phloem and a surrounding layer of the future pericycle cells. In this way the basic organization of the radial pattern of tissue layers is completed before the heart stage. Only in the torpedo-shaped embryo is the radial pattern completed: the ground tissue divides by periclinal cell divisions into an outer cortex layer and an inner endodermis layer.

The transition from the late globular stage into the heart stage occurs rather quickly. In the apical region cells start to divide faster at two different sites, from which the primordia of the cotyledons are to emerge. On the apical side, the embryo flattens. Further cell divisions lead to the growth of the cotyledon primordia and the embryo becomes heart shaped. The basic plant body begins to take shape as the embryo makes the transition from radial to bilateral symmetry. At this point, the early phase of embryonic development is over.

During the early heart stage, the embryo consists of about 250 cells. As cells in the cotyledon primordia and the body axis divide, the embryo grows in length. From the mid-torpedo stage, the cotyledon primordia grow by following the bends of the ovule. As a result, their tips lie at the same height as the root. In the body axis the cell files increase in size by both intercalary growth and cell division in the root meristem (section 6.1.3). The individual tissue layers now have visibly different morphologies because they contain different cell types. Only at the end of embryo development does the shoot meristem become visible as a bulge between the inner sides of the cotyledons (section 6.1.4).

There are three major phases of embryo development. During early embryogenesis, pattern formation generates the basic body plan. This phase finishes when the embryo becomes heart shaped. At this point, the primordia of seedling structures become visible. In the second phase the primordia develop, growing by cell division and being further differentiated. The final phase is characterized by physiological changes that prepare the embryo for seed dormancy (section 5.6.1).

6.1.1 The emergence of the basic body plan: pattern formation during early embryogenesis

The seedlings of flowering plants have a remarkably uniform structural organization, which can be thought of as a superimposition of two patterns (Figure **6.2**). Along the longitudinal axis (which becomes the main axis of the adult plant) an apical–basal pattern forms from the following elements, listed from top to bottom: shoot meristem, cotyledons, hypocotyl, embryonic root, root meristem, and central root cap. Perpendicular to this axis a radial pattern forms from the concentric layers of all the main tissue types and their derivatives, which, from outside to inside, are: the epidermis (which in the root is surrounded by the lateral root cap), the ground tissue from which the cortex and endodermis derive, and the vascular tissue in the centre (the central cylinder) with its pericycle, phloem, and xylem.

The body organization of the seedling is clearly derived from the basic body plan of the heart-shaped embryo. The embryo is divided from apex to base into three discrete regions: apical, central, and basal. Radial division is nearly complete: the epidermis, ground tissue, and pericycle surround the precursors of the phloem and xylem. Using clonal analysis, it has been found that the tissues of the seedling emerge from their primordia in the heart-shaped embryo, although there is no clonal relationship between the regions of the embryo and the apical–basal pattern elements of the seedling. The apical region produces the shoot meristem and the main part of the cotyledons. However, a small part of the cotyledons (the *shoulder region*) derives

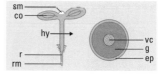

Figure 6.2 Body organization of the seedling. Two patterns can be distinguished: an apical–basal pattern along the body axis which consists of the shoot meristem (sm), cotyledons (co), hypocotyl (hy), root (r), and root meristem (rm). Perpendicular to the axis, a radial pattern consists of the concentrically arranged main tissues: epidermis (ep), ground tissue (g), and vascular tissue (vc). (According to Mayer *et al.* 1991.)

from the central region, which is otherwise concerned with constructing the hypocotyl of the embryonic root and the initials of the root meristem. The basal region produces the rest of the root meristem, the quiescent centre, and the initials of the central root cap. It seems that the origin of the cells is not important in the formation of the apical–basal pattern. However, the regions that are derived from defined cell groups in the early embryo are still important for apical–basal pattern formation, as can be seen in pattern mutants (see below).

In *Arabidopsis*, cell divisions during early embryogenesis are very regular, so its basic body plan can be traced back to certain cell groups in the early embryo. However, this does not mean that a cell is committed to a specific fate at an early stage and that this fate is passed on to its progeny. As an example of this, *fass* embryos have very irregular early cell divisions, and the characteristic cell shapes and cell groups of the primordia cannot be recognized. Despite this, all the structures of the seedling eventually form in the right places, but they are deformed. The *FASS* (*FS*) gene is required for morphogenesis, but not for pattern formation. In the embryos of other plant species that have irregular early cell divisions, the characteristic structures nevertheless emerge. Furthermore, zygotic and somatic embryos of one species can generate the same basic body plan, but with very different cell divisions. These observations show that pattern formation in *Arabidopsis* normally leads to a regular cell division pattern. However, this is a result of the pattern formation process, and not part of it. It is more likely that a cell's developmental fate is determined by its interaction with and its position relative to its neighbouring cells. That this is the case has been indicated by investigations with embryonic pattern mutants of *Arabidopsis*.

In the pattern mutants, either the apical–basal pattern or the radial pattern of the seedling is altered (Table **6.1**). Mutations in four genes, *GNOM* (*GN*), *MONOPTEROS* (*MP*), *FACKEL* (*FK*), and *GURKE* (*GK*) change the apical–basal pattern of the seedling in a characteristic fashion without disturbing its radial pattern (Figure **6.3**). The changes can be seen in the basic body plan of the mutant embryos, and are therefore probably the result of disruptions in pattern formation in the young embryo. The most drastic changes can be seen in *gn* seedlings: in the most extreme case, these become ball shaped and have no regional differences along their apical–basal axis. Most often, though, the apical end is reduced in size to varying extents, but their root is always absent. This phenotype can be traced all the way back to the zygote—compared to the wild type, the *gn* zygote elongates less, and divides almost symmetrically. The enlarged apical daughter cell divides either tangentially or horizontally, instead of longitudinally. Its two daughter cells divide like the apical cells of a normal embryo, leading to a double octant embryo. The basal daughter cell of the *gn* zygote is smaller. It only divides horizontally a couple of times, leading to a shortened suspensor. The uppermost cell, which abuts the enlarged octant embryo does not behave like an hypophysis—it does not go through a predefined division programme that would result in the formation of a basal region. In tissue culture, *gn* seedlings whose bottom ends have been removed are incapable of forming new roots, and can only form calli. All these results suggest that the *GN* gene product is important for determining the embryo's apical–basal polarity, on which the partitioning of the longitudinal axis into different regions depends.

The phenotypes of the other apical–basal pattern mutants provide some clues as to the development of the regions. The *mp* seedlings have no hypocotyl, root, or root meristem. This is because of defects at the octant stage: as with *gn* embryos, the *mp* octant pro-embryo has two extra cell layers, yet the suspensor appears normal. During the globular stage, the apical region develops normally while the cells of the central region divide irregularly. The cell that borders the pro-embryo undergoes repeated horizontal divisions, in

Gene	Mutant phenotype
GNOM (GN)	Division of the zygote (more or less symmetrically); the entire apical–basal pattern is affected
MONOPTEROS (MP)	Instead of 8, there are 16 cells in the pro-embryo; the basal part of the embryo develops abnormally
FACKEL (FK)	Central region in the globular-stage embryo is abnormal; the seedling lacks a hypocotyl
GURKE (GK)	Apical region abnormal; no cotyledon primordia and no shoot meristem.
HOBBIT (HBT)	Abnormal cell division of the hypophysis in the globular embryo; no root meristem activity
SHORT ROOTS (SHR)	No endodermis in the pro-embryo; seedling has a short root
SCARECROW (SCR)	No separate endodermal and cortical layers

Table 6.1 Genes for pattern formation in the *Arabidopsis* embryo.

Figure 6.3 Apical–basal pattern mutants. Top row: phenotypes of mutant seedlings; co, cotyledon; hy, hypocotyl; r, root. Middle row: early stages of mutant embryos that differ from the wild type. Bottom row: the corresponding wild-type embryo stages. ac, bc, apical and basal daughter cells of the zygote. (According to Jürgens *et al.* 1994.)

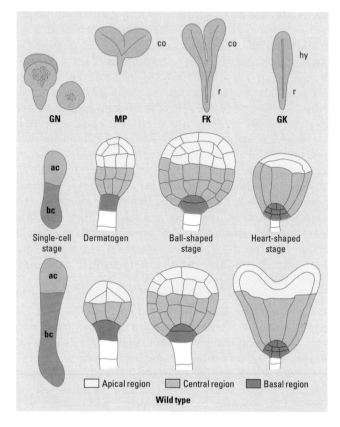

a similar fashion to the suspensor cells. All the above leads to the lack of hypocotyl, root, and root meristem in *mp* seedlings. Moreover, the primary change in the *mp* octant pro-embryo suggests that the function of the hypophysis as the origin of the basal region in the embryo is normally induced by a signal from the pro-embryo. This result has been supported by experiments using tissue culture: *mp* seedlings that have been wounded or dissected can still form a root. The *MP* gene is therefore not necessary for root formation, although *mp* embryos lack a root.

The phenotype of *gk* seedlings is complementary to that of *mp*: the apical structures, shoot meristem, and cotyledons are all affected. The *gk* phenotype becomes visible when, at the transition to the heart stage, the apical end of the embryo remains round instead of showing the protrusions that mark the cotyledon primordia. Therefore the complementary phenotypes of *mp* and *gk* mutants suggest that the cotyledon primordia are initiated in the apical region of the embryo, while neighbouring cells in the central region are possibly prompted to participate in the production of cotyledons by a signal from the apical region.

In *fk* seedlings, the cotyledons seem to be connected directly to the root, and the hypocotyl is missing. Their phenotype becomes visible in the early globular stage when cells in the central region do not divide asymmetrically. It seems as if the hypocotyl primordium is already genetically defined before it can be distinguished morphologically from the root primordium. However, contrary to *mp* embryos, *fk* embryos have a basal region from which the quiescent centre of the root meristem is constructed. This leads us to assume that the basal region normally induces adjacent cells of the central region to become the initials of the root meristem. These initials will then produce most of the embryonic root. The inference that the basal region induces the initials of the root meristem is also supported by the phenotype of mutations in the *HOBBIT* (*HBT*) gene. In *hbt* embryos, the hypophysis does not divide properly and later on the portion of the embryonic root that is normally made by the initials of the root meristem fails to form.

Since the phenotypes of the apical–basal pattern mutants define three regions along the longitudinal axis—apical (*gk*), central (*mp* and *fk*), and basal (*mp*)—it is assumed that these three regions are essential for the formation of the apical–basal pattern. The emergence of these three regions depends on events in the zygote that are not yet understood, and which lead to asymmetric cell divisions and indirectly to an apical–basal polarity of the embryo. The *GN* gene seems to participate in this. That the *GN* gene is prerequisite for manifestation of the effects of the other genes (which are all more regionally specific) could be shown by phenotypic analysis of double mutants. *Gn* is epistatic to *mp*: this means that when the *GN* gene is inactive, the *MP* gene cannot have an effect.

A radial pattern emerges from a series of periclinal divisions, starting at the periphery and ending in the centre of the embryo. The dividing cells of the octant-stage embryo give the outer epidermal primordium and inner cells. Subsequent cell divisions in the epidermal primordium are strictly anticlinal, thus maintaining the integrity of the layer. The expression pattern of the *ARABIDOPSIS THALIANA MERISTEM LAYER 1* (*ATML1*) gene, which encodes a homeodomain-containing putative transcription factor, visualizes the formation of the epidermal primordium. Expression of the *ATML1* gene starts in the apical daughter cell of the zygote and becomes restricted to the epidermis primordium following the periclinal divisions of the eight cells of the octant-stage embryo. Subsequently, the *ATML1* gene continues to be expressed in the outer (epidermal) layer of the embryo, the L1 layer of the shoot meristem, and the epidermis of developing leaves and floral organs. Fairly soon after its formation, the epidermal primordium expresses another gene, the *LIPID TRANSFER PROTEIN* (*LTP*) gene, which also continues to be expressed in the derivatives of the epidermal primordium throughout development. Mutations in several genes affect the radial pattern. In most cases, however, the mutant effects are indirect. For example, mutations in the *KNOLLE* (*KN*) gene primarily affect cytokinesis (cell division), resulting in an incomplete physical separation of the daughter cells. In the *kn* octant embryo, the primordium of the epidermis is not separated from the inner cells by periclinal divisions, and consequently the epidermis-specific *LTP* gene is expressed ubiquitously throughout the *kn* globular embryo. Later on, though, the *LTP* gene is no longer expressed in the middle of the embryo. Here, the vascular system develops. These finding show that there must be a large degree of flexibility in the emergence of the radial pattern.

Double mutant analysis can be helpful in distinguishing between genes that play specific roles in radial pattern formation and other genes that are merely required for the normal development of the radial pattern. For example, *short root* (*shr*) embryos do not develop an endodermis, and *wooden leg* (*wol*) embryos have fewer cells than normal in the vascular system, and those cells that are present develop only into xylem. Investigations of double mutants with *fass* (*fs*) have shown that the primary defects are different: the *shr* phenotype is not suppressed by *fs* although *shr fs* embryos produce more layers of ground tissue by abnormal cell divisions. In contrast, the defects in the vascular tissue of *wol* embryos can be completely compensated for by extra cell divisions in the *wol fs* double mutant. Therefore, the precise pattern of periclinal cell divisions is not essential for the formation of different tissue layers. Instead, it is assumed that local interactions between cells are responsible.

As the last step in radial pattern formation, the outer cortex and the inner endodermis originate from asymmetric cells divisions in the ground tissue. As discussed above, *shr* embryos lack the endodermis, regardless of the number of ground tissue layers. This result suggests that the *SHR* gene confers endodermal cell identity. Mutations in another gene, *SCARECROW* (*SCR*), which encodes a putative transcription factor, also inhibit the asymmetric cell divisions. In *scr* embryos, however, the mutant cell layer displays both endodermal and cortical traits. This defect is suppressed in the *scr fs* double mutant, which has supernumerary cell layers, suggesting that the *SCR* gene does not determine cell fate but may play a role in the division process. This is consistent with the expression of the *SCR* gene in the ground tissue before asymmetric cell divisions, while subsequent expression is restricted to the endodermal cell layer.

6.1.2 The growth and differentiation phase

From the heart stage up until the mature embryo, the regions and tissues that were established earlier are further subdivided. During this time, the subdivision of the axis into hypocotyl and root primordia is striking, as the now-active primary root meristem forms the largest part of the embryonic root (section 6.1.3). In contrast, the shoot meristem is wholly indistinct, producing no morphologically recognizable structures. The cotyledon primordia grow in length significantly; their vascular strands are formed in addition to the epidermis and ground tissues already present.

The radial patterns of the hypocotyl and the root are completed. The ground tissue differentiates into the outer cortex and the inner endodermis that surrounds the vascular cylinder. Additionally, in the root the epidermis and the lateral region of the root cap emerge from common initials in the root meristem (section 6.1.3).

6.1.3 The origin of the primary root meristem and the embryonic root

According to histological studies, the primary root meristem is composed of two parts that originated in different parts of the embryo (Figure **6.4**). The quiescent centre of the root meristem and the initial cells of the central root cap are derived from the hypophysis and are therefore originally derived from the basal daughter cell of the zygote. In contrast, the initials of the root meristem, which form the cell files that grow upwards and thus contribute to the formation of the embryonic root, derive from the lower tier of the octant pro-embryo. Thus they originate from the apical daughter cell of the zygote. These results have been confirmed with clonal analysis of root development in the

Figure 6.4 Origin of the root and the root meristem. The quiescent centre (qc) of the root meristem (rm) and the initials of the central portion of the root cap (irc) are derived from the hypophysis (hp) and therefore from the basal daughter cell (bc) of the zygote. The initials of the cell files of the root (ri) have their origins in the lower tier of the octant pro-embryo (lt), which has its origin in the apical daughter cell (ac) of the zygote. The upper region of the root (ur) is not formed by the initials (ri). ut, Upper tier of the octant pro-embryo; su, suspensor; hy, hypocotyl; r, the part of the root that is formed by the meristem. (According to Scheres *et al.* 1994.)

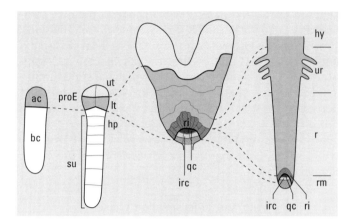

embryo of *Arabidopsis*. This clonal analysis supports the view, based on phenotypes of apical–basal pattern mutants (section 6.1.1), that local interactions between cells play a role in the formation of the root meristem.

The root meristem becomes active at the heart stage. The initials of the central root cap produce four layers of cells towards the bottom, until the end of embryonic development. The initials on top of the quiescent centre produce cell files that form the largest part of the embryonic root. The primary root meristem has a characteristic radial organization that is reflected in the concentric arrangement of the root tissues (Figure **6.5**). Eight cell files each are found in the cortical layer and the endodermis. These numbers are constant. Each cell file of the cortex and the endodermis derives from a common initial in the root meristem. There are approximately 16 cell files in the epidermis and twice as many in the lateral root cap. Their numbers tend to vary a little. However, a clonal analysis has shown that there is a common initial in the root meristem for one cell file in the epidermis and two cell files in the lateral root cap underneath the endodermis. Approximately 12 cell files of the pericycle surround the phloem and the xylem. Although these cell numbers may suggest that the fates of the initials in the root meristem are determined in the late globular embryo, there is evidence, from ablation studies on seedlings, for position-dependent cell fate choices in the root (section 6.2.1).

The part of the embryonic root emerging as a result of meristematic activity is covered by the lateral root cap. The adjacent part that lacks the root cap contains the 'collet' that designates the transition from root to hypocotyl. Using

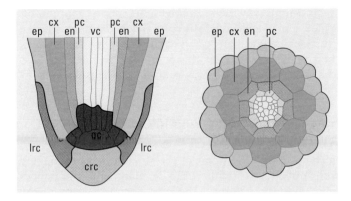

Figure 6.5 Organization of the root meristem and the primary root. The root meristem (left, in bold outline) is composed of the quiescent centre (qc), and the surrounding initials of the cell files of the root tissue, and of the central root cap (crc). Common initials are present for the lateral root cap (lrc) and the root epidermis (ep) as well as for derivatives of the ground tissue (cortex (cx) and endodermis (en)). To the right the concentric arrangement of the root tissues is shown; vc, vascular cylinder; pc, pericycle. (According to Dolan *et al.* 1993.)

clonal analysis, it has been shown that the upper part of the embryonic root derives from the cells of the central region, which, in the globular stage, are adjacent to the future initials of the root meristem (Figure **6.4**). There is no sharp boundary between these two parts, and the clones often end above or below this theoretical border. Therefore, the fate of the cells does not depend on their origin.

6.1.4 The origin of the primary shoot meristem

In *Arabidopsis* embryos, the primary shoot meristem is nearly inactive. It is composed of only a few cells, and so it is difficult to distinguish before the non-meristematic cells differentiate in the apical region of the embryo. Only at the end of embryogenesis does a small hump of cells, rich in cytoplasm, appear between the bases of the cotyledons.

It is not known when and how the shoot meristem develops as a functional unit. Most probably, the apical region is partitioned into the central area that forms the shoot meristem and the cotyledon precursors at the sides (sections 6.1.1 and 6.2.5). This suggestion is supported by the expression patterns described recently for two genes. The shoot meristem-specific *STM* gene (section 6.2.5) is expressed in the presumptive shoot meristem from the late globular stage, and the *ANT* gene is expressed in two flanking regions from which the cotyledon primordia are to develop. Histological studies on successively younger stages have helped identify and locate the shoot meristem precursor cells in the apical pole of the embryo during the early heart stage. However, this does not mean that these cells are already destined to become the shoot meristem. At this point there are two distinguishable layers: an outer layer that derives from the embryonic epidermis, and a subepidermal layer that shortly afterwards divides to form two layers. In this way the presumed shoot meristem primordium develops the layered structure that is characteristic of a functional shoot meristem (section 6.2.5). In the embryo, the vascular tissue abuts this group of cells from beneath. The vascular tissue develops out from the central region.

6.2 Postembryonic vegetative development

Postembryogenic (vegetative) development has its origins in the primary meristems of the shoot and root. Looking from outside, postembryonic development commences with the germination of the seed, when the root of the seedling begins to grow. But if the pattern of cell division is taken as a reference point, the transition from embryonic to postembryonic development can be considered to be independent of germination. The intercalary growth that is characteristic of the embryo is, in many species, supplemented during late embryogenesis by localized (meristematic) growth, when the primary root meristem produces parts of the seedling's root. In comparison, the degree of development of the primary shoot meristem of embryos of different plant species varies. For example, the shoot meristem of the maize embryo has already produced several leaf primordia before it becomes dormant, but mature *Arabidopsis* embryos have no well-formed leaf primordia. However, groups of founder cells for primary leaves are already established, as can be seen by clonal analysis of the primary shoot meristem (section 6.2.5). Furthermore, some *Arabidopsis* mutants produce primordia of primary leaves while they are still embryos.

6.2.1 Development of the primary root

The primary root meristem of the seedling is a very regular structure (*see* Figure **6.5**). The quiescent centre, four cells that do not divide, is surrounded by the initials, which are actively dividing stem cells. Towards the bottom, a layer of initials constantly produces new cells for the central part of the root cap, which is eroded away continuously as it spears its way through the soil. Towards the top, a layer of initials produces files of cells that elongate the already existing root.

The primary root grows mainly by cell division within its meristematic zone. Newly produced cells elongate the root that was built during embryogenesis. When cells leave the meristem they are incorporated into the zone of elongation. Here they elongate along the longitudinal axis of the root. Cell elongation also contributes to root growth. Finally, the cells enter the zone of differentiation, where they begin to produce the features that will later characterize them. In each layer of tissue, specific cell types develop; one example is the root hair in the root epidermis (section 6.2.2). The cells of the endodermis produce the Casparian strip, which separates the vascular bundle from the surrounding cortical root tissue. The pericycle, from which lateral roots originate (section 6.2.3), surrounds the vascular tissue, which by now is differentiating into clearly distinguishable phloem and xylem strands.

Different mutations can alter the primary root of *Arabidopsis*. In extreme cases, the root corresponds to the embryonic root, because the meristem makes no cells in the postembryonic phase. Other mutants begin to show their phenotypes when newly produced cells enter the zone of elongation: depending on which particular gene is mutated, the cells of different tissues can swell abnormally. It would seem as if the cells are unable to elongate. There are also mutants that have tissue-specific defects. The *short root* (*shr*) mutant has no endodermis, and thus no Casparian strip. These mutants do not only show their defects postembryonically, the embryos also lack the same tissue, not only in the root but also in the hypocotyl. Therefore, these seem to be radial-pattern mutants (section 6.1.1). Since the defects also occur in the postembryonically constructed part of the root, it would seem that the mutated genes are needed in the embryo as well as in the root meristem. Alternatively, the mutants do not establish root meristem initials in the embryo. In both cases, embryonic pattern formation determines directly or indirectly the organization of the primary root.

Cell ablation experiments have recently provided evidence that the initials of the root meristem lack intrinsic information about the kinds of cell they are to produce. If a cortex/endodermis initial is eliminated, its place is occupied by two new cells derived from the underlying pericycle initial. These new initials then produce cell files, resulting in the formation of nine, instead of eight, cortical and endodermal cell files in the root. Similar experiments performed on daughter cells of the initials indicated that the mature tissues of the root have an instructive influence on the fate of the newly formed cells. Thus, the root meristem initials may best be viewed as a group of 'naive' stem cells which may be kept mitotically active by the adjacent quiescent centre of the root meristem.

6.2.2 Development of an epidermal cell type in the root: the root hair

The root epidermis consists of two types of cells: hair cells and non-hair cells. Root hair cells are initially shorter and make a local hump close to the basal cell wall, and from there grow outwards as long tubes (the root hairs). The non-hair cells are longer, have no hump, and are coated with deposits.

In *Arabidopsis*, these two cell types occur in different cell files in the epidermis: the eight-cell files that overlie each border between two neighbouring cortex cell files consist only of root hair cells; the other cell files consist only of non-hair cells. How is the fate of the cells determined in the root epidermis (pattern formation)? Even though the uniform cell files would suggest that initials in the meristem have predetermined fates, and that these fates are passed to their files of daughter cells as though they were clones, there is some evidence that, instead, interactions between the cell files determine the cells' fates.

If the epidermis is detached from the layers of root beneath, additional cells differentiate to form root hairs. It seems that the root hair cell is the ground state of differentiation in the root epidermis. Mutations in the *CONSTITUTIVE TRIPLE RESPONSE 1* (*CTR1*) gene increase the numbers of root hair cells by about 30 per cent, and mutant seedlings also possess root hair cells above the cells of the cortex. Since the *CTR1* gene regulates negatively the response to ethylene, ethylene seems to promote the production of root hair cells. The product of the *CTR1* gene, for example following a release of ethylene, could only be inactivated at points where two neighbouring cortex cells abutted one another (Figure **6.6**). Mutations in another gene, *TRANSPARENT TESTA GLABRA* (*TTG*) also lead to the formation of root hairs from cells that otherwise would not differentiate this way. The *ttg* phenotype is suppressed by a constitutive expression of the *R* gene from maize, which itself prevents the formation of the hair cells. The interaction of *ttg* with the *R* gene is similar to the process involved in the production of trichome cells in leaves (section 6.2.7). Therefore we can also assume that local interactions between cells in the root epidermis determine whether a cell will be a hair cell or a non-hair cell. However, these supposed interactions do not have any effect on cells within a cell file, because all cells in one cell file are either root hair cells or not. Therefore one could postulate that the cell's fate is always determined within a certain distance of

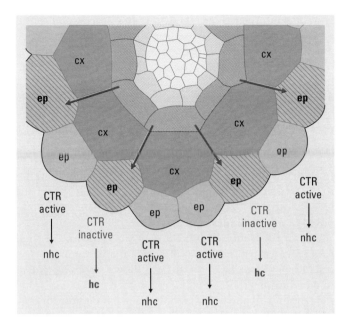

Figure 6.6 Model of the formation of hair cells in the root epidermis. Hair cells (hc) emerge in the cell files of the root epidermis (ep) that lie above the cell wall between two cortex cell files. The other epidermal cells differentiate to form non-hair cells (nhc). It has been postulated that a signal (arrow) coming from inside the root (for example, ethylene) inactivates a negative regulator in the presumptive hair cells (for example, the *CTR* product), and in this way, allows differentiation.

the meristem (for example within the zone of elongation), and from then on is fixed.

How does a root hair cell get its shape? This particular cell morphogenesis has been well studied. At the basal end, close to the cell wall, a hump is produced that is close to the cell wall of the neighbouring next-youngest cell. This hump then grows. The self-elongating tube grows by deposition of material at its tip, in a similar fashion to the pollen tube. *Root hair-defective* (*rhd*) mutants have been isolated in which root hair formation is disturbed: for example, the hump might not grow, or in a different mutant it might branch and form several tubes, or the tubes might not grow straight. These mutants form the raw material for defining the genetic steps of root hair morphogenesis.

6.2.3 Lateral and adventitious roots

Primordia of lateral roots do not appear as a defined cell group set aside from the primary root meristem. Moreover, lateral roots emerge from the pericycle cells which begin to divide at quite some distance from the root tip. Not all pericycle cells initiate lateral root growth; only those that lie close to the xylem strands do this. In *Arabidopsis*, approximately 14 neighbouring pericycle cells make up the lateral root foundation which becomes a primordium before the secondary root meristem appears. The growing lateral root breaks through the adjacent cell layers, and then forces its way through the epidermis. At the tip of the lateral root is a meristem, with an organization similar to that of the primary root meristem. However, analysis of the expression of tissue-specific markers suggests that the tissue organization of the lateral root is well established before the meristem of the lateral root appears. In this sense, the development of the lateral root appears to recapitulate the development of the primary root in the embryo (sections 6.1.3 and 6.2.1). In the secondary root meristem, there are also initials that produce cell files, which make up the concentric organization of different tissue layers. One difference to the primary root is that the number of cell files per tissue layer is variable. For example, the cortex layer can be composed of 6–11 cell files instead of eight.

Lateral roots differ from the primary root not in their general construction, but in their origin and the number of cell files per tissue layer. In this way, lateral roots are similar to adventitious roots which emerge by regeneration from, for example, hypocotyls that have been cut off or from calli. Although it is still not known how the foundations of lateral and adventitious roots emerge, the newly formed root meristems seem to be similarly organized and also subject to the same genetic regulatory mechanisms as the primary root meristem. For example, mutations that forestall the activity of the primary root meristem have the same effect on the secondary root meristem. Additionally, the same cell-type-specific defects have been observed in both the primary root and the lateral roots in other mutants.

6.2.4 Shoot development

Arabidopsis is a rosette plant. Its leaves emerge short distances from each other, with no significant growth in internode length. After flower induction, the shoot grows and the leaves at the bottom of the shoot are arranged into a rosette. In the axils of the rosette leaves there are foundations for side shoots (secondary shoot meristems) that usually grow late, or not at all. The rosette leaves have a characteristic arrangement relative to one another. This is called *phyllotaxis*. The first-formed *primary leaves* grow opposite each other, forming nearly at right angles to the cotyledons. The slight deviation from 90° starts the screw-like arrangement of the following leaves, which are set at 137° intervals to each other. Which way the leaves turn is left to chance, according to the way the axis of the primary leaves tips in relation to the cotyledons. Until the onset of flowering during long days, 8–9 leaves are formed, and

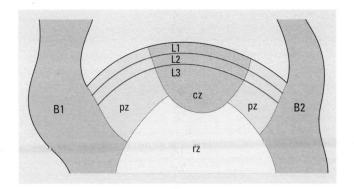

Figure 6.7 Functional organization of the shoot meristem. The shoot meristem is composed of layers (L1–L3) and zones (cz, central zone; pz, peripheral zone; and rz, rib zone). Each zone contains all the layers. The L1 layer forms the epidermis; L2 and L3 form the subepidermal tissues of shoot, leaves, and flowers. The central zone is a group of non-permanent stem cells. The peripheral zone gives rise to leaf and flower primordia. The rib zone contributes to shoot growth. B1, B2, leaf primordia. (According to Steeves and Sussex 1989.)

then the shoot grows. If, because of short-day conditions, the onset of flowering is delayed, more rosette leaves form. In contrast, early flowering mutants produce either fewer, or in extreme cases, no rosette leaves at all. How the aerial parts of the vegetative plant form can be seen in investigations of the primary shoot meristem.

6.2.5 The shoot meristem is organized in layers and zones

The primary shoot meristem of the seedling is characteristically organized into layers and zones (Figure **6.7**). The outer (L1) layer of three emerges from the epidermis of the embryo. Later, it will form the epidermal tissue of the shoot, the leaves, and the flower organs. The integrity of the L1 layer is secured by anticlinal cell divisions. Only seldomly is the cell division in L1 oriented abnormally so that a daughter cell will come to lie towards the inside. This cell will lose its epidermal character and its daughter cells will produce subepidermal structures. The cells of the inner layers of the shoot meristem emerge from subepidermal cells within the embryo. The cells of the L2 layer also divide anticlinally. They produce subepidermal structures—for example the sporogenic tissues of the flower that produce the germ cells. Beneath the L1 and L2 layers (the *tunica*) there is the L3, or *corpus* layer. The cells of the corpus divide anticlinally *and* periclinally. Unlike the well-defined border between the epidermis and subepidermis that originates from the L1 and L2 layers in the meristem, the origin of the subepidermal tissue is variable and is not clearly L2 or L3.

The shoot meristem is further subdivided into several zones, each fulfilling a different function. A central zone at the tip contains actively dividing cells which continuously renew the meristem, producing daughter cells to the side and bottom. The central zone is also important for the

---

Box 6.1 Clonal analysis

Clonal analysis helps to identify which parts of the adult organism are derived from which particular cells from the early developmental stages. For example, if the offspring of a given cell in a meristem differentiate only into epidermal structures, one can conclude that the developmental potential of that cell was limited. However, if descendants of a particular cell can be found in different leaves, the initial cell was not part of a single leaf primordium. Analysis of several clones, induced at a particular developmental stage, can give a picture of the meristematic organization, that generates the structures of the adult plant.

Clones can be induced by several methods. Using radiation, or chemical mutagenesis, homozygous mutant cells can be produced in heterozygous mutant embryos. The offspring of these cells have a distinguishable phenotype that can be used as a marker. For parts of the plant that grow above ground, pigments (such as *albino* and *fusca*) are most useful as markers. In the descendant cells of L2 or L3, these markers lead to either a lack of pigments (*albino*), visible as white spots on the leaves and stem; or to accumulations of anthocyanin (*fusca*), seen as dark purple spots. Frequencies of these clones are so low that, in general, the marked areas comprise descendants of single mutant cells. For parts of the plant that grow from the root meristem, and that remain underground, different markers are necessary. One example is colchicine-induced polyploidy, which can be seen in mitotic preparations. Nowadays, transgenic plants containing a 35S::GUS fusion gene are used, the ubiquitous expression of which is inhibited by the Ac transposon from maize. This transposon is integrated between the 35S promoter and the coding region of *B*-glucuronidase (GUS). In some cells of the developing offspring of such plants, the transposon can excise itself from the fusion gene, albeit infrequently. The effect is that in these cells and their descendants (clones) the GUS protein is expressed and the enzymatically active cells can be displayed using a colour reaction.

integrity of the shoot meristem. It is flanked by the peripheral zone where the leaves and axillary meristems form. The leaf precursors are replaced by new cells from the central zone and leave the meristem. The subepidermal cells divide periclinally, so that the leaf precursor cells form leaf primordia. The axillary meristems they contain remain inactive as long as they are inhibited by the central zone of the primary shoot meristem (*apical dominance*). Beneath the central zone is the rib zone. Its cells contribute to the growth of the shoot.

In the mature embryo, the shoot meristem contains about 100 cells and is approximately 50 μm in diameter. How this meristem produces vegetative structures has been investigated using clonal analysis of the subepidermal tissue (L2 and L3; Box **6.1**). In order to carry out this investigation, mature seeds were treated with X-rays. Flowering plants were grown from the treated seeds, and white spots appearing on the rosette leaves, shoots, and green parts of the flowers were evaluated. The clones differed in their

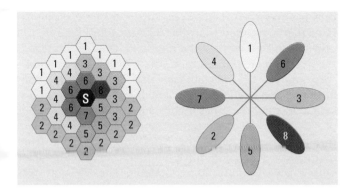

Figure 6.8 Fate map of the shoot meristem of the mature embryo. A fate map has been deduced from the frequency and distribution of clones in the L2 layer of rosette leaves and flowers. This map gives a probability (depending on location) that a cell will contribute to a given structure in the shoot meristem. The numbers on the left in the diagram relate to leaves of the rosette (1–8) that are formed one after the other before the induction of flower formation. In the centre of the fate map there is a cell (s) that has the highest probability of forming the shoot and the flowers. (According to Irish and Sussex 1992.)

sizes and locations. There were no clonal restrictions that could show limitations of the developmental potential of individual cells within the shoot meristem (Figure **6.8**). Some clones were restricted to single leaves, but other clones stretched over neighbouring rosette leaves. Only a few clones extended from the rosette leaves up to the inflorescence. Clones on side shoots were accompanied by clonal spots in the centre of the leaves from whose axils they had grown; the axillary meristem therefore arises from parts of the leaf primordium. The two primary leaves had more, smaller, clones than the following leaves, suggesting that their foundations had embryonic origins. Data collected from these experiments enabled the construction of a model for the organization of the shoot meristem of the mature embryo (Figure **6.8**).

The model predicts that certain cells in the L2 (and L3) layer contribute to certain structures, according to their location in the meristem, but that these cells do not have a predetermined fate. The point at which their fates are determined is possibly when the leaf primordium becomes distinguishable from the meristem. Until then, the cells of the foundation keep dividing. The peripheral cells around the meristem divide very little before they contribute to the primary leaves. The cells in the middle of the meristem that will produce future leaves and shoot contribute in a variable way to the formation of these structures. Their flexibility is shown by the way that clones in the shoot can push other cells out of the way so that they can exclusively produce the flowers. We can conclude from this that the fate of a cell depends mainly on the information that it exchanges with its neighbouring cells. The organization of the shoot meristem into zones appears to be maintained by signals that the cells pass between themselves, ensuring that cell populations that are parted from the side are substituted by cell divisions in the centre of the meristem. Only in this way is the size of the meristem kept constant over a given period.

To analyse the mechanisms that produce the shoot meristem and regulate its size and organization, it is necessary to identify the genes involved in the process and their products. One of the essential genes for shoot meristem formation seems to be the *KNOTTED (KN)* gene from maize. This gene codes for a putative transcription factor with a DNA-binding homeodomain. The *KN* gene was identified because dominant mutant alleles cause knotted deformations of the leaves. This phenotype is caused by ectopic expression of the *KN* gene in the developing leaves. Constitutive expression of *KN* cDNA in transgenic tobacco can cause a shoot meristem to form on a leaf. Normally the *KN* gene is expressed in the shoot meristem of maize but not in the founder cells of the leaf (in the peripheral zone of the meristem). Therefore, the KN protein seems to promote meristematic activity in cells. However, there are no loss of function alleles of *KN*, so it is not clear what biological consequences the lack of KN protein might have for the meristem. In *Arabidopsis*, recessive mutations in the *KN*-related gene *SHOOT MERISTEMLESS (STM)* lead to a defective shoot meristem. The *STM* gene has a similar homeodomain to that of *KN*. The *stm* embryo has only a rudimentary shoot meristem from which no leaves are produced after germination; later, leaves emerge from the axils of the cotyledons, but the shoot is absent (Table **6.2**). It can be assumed that genes like *KN* and *STM* have similar roles in other plant species for the origin and maintenance of the shoot meristem. Another gene, *ZWILLE (ZLL)*, which is necessary for the formation of the shoot meristem in the embryo, has been identified in *Arabidopsis*, although it is not needed for the production of adventitiously formed shoot meristems. Mutant *zll* seedlings do not form leaves; the apical cells between the cotyledons seem not to be meristematic, and instead are differentiated. Approximately 2 weeks later, *zll* seedlings are able to form adventitious meristems in the axils of cotyledons which go on to form fertile flowering shoots.

Mutations in other genes in *Arabidopsis* change the organization of the shoot meristem from the embryo (Table **6.2**). Both the primary and secondary shoot meristems (including the flower meristem) can be affected. Mutations in the *WUSCHEL (WUS)* gene eliminate structures that would normally be formed by the central zones of the shoot and

Gene	Mutant phenotype
SHOOT MERISTEMLESS (STM)	Defects in formation of the shoot meristem
ZWILLE (ZLL)	No shoot meristem formed in the embryo
PINHEAD (PNH)	No shoot meristem formed in the embryo
WUSCHEL (WUS)	No central zone in the shoot meristem; central defect in the flower
CLAVATA 1 (CLV1)	Shoot meristem enlarged; additional pistil in the centre of the flower and further outer flower organs
ALTERED MERISTEM PROGRAM 1 (AMP1)	Additional cotyledons, abnormal phyllotaxis
EMBRYONIC FLOWER (EMF)	No vegetative phase: instead of leaves, the seedling forms flowers
EXTRA COTYLEDON (XTC)	Shoot meristem in the embryo forms additional cotyledons
FUSCA 3 (FUS3)	Shoot meristem in the embryo forms primary leaves

Table 6.2 Genes for the development of the shoot meristem of *Arabidopsis*.

flower meristems. Mutant *wus* seedlings produce primary leaves only. Later, defective secondary meristems, which can themselves only produce two leaves, emerge, until eventually, instead of a rosette, a chaotic bunch of leaves grows. Sometimes *wus* plants flower; a shoot develops late, and instead of a single cauline leaf a bundle of leaves forms. Out of the axil of a cauline leaf, a shoot with a bunch of similar leaves can grow, and so on. At the end of development, an almost empty flower forms at the top of the shoot. Its pistil is missing; in its place is a single solitary stamen, surrounded by the normal number of petals and sepals. The *WUS* gene appears to be important for the formation of the central zone. If this zone is missing, meristematic tissues cannot renew themselves, and there is no apical dominance, so the formation of secondary meristems is not suppressed. A *wus* complementary phenotype is caused by mutations in three *CLAVATA* (*CLV*) genes in *Arabidopsis*: the central zone of the shoot meristem is enlarged, leading to abnormal phyllotaxis. The shoot itself can split (*fasciation*), and the flower meristem is also enlarged. More elements are constructed in the organ whorls, and the centre of the flower contains the extra whorl of an additional pistil. The *CLV1* gene encodes a transmembrane receptor Ser/Thr kinase which may be involved in cell–cell signalling within the shoot meristem. It can be assumed that the effect of the *WUS* gene product is limited to the centre of the shoot meristem by the products of the *CLV* genes, and that these interactions determine the size of the shoot meristem.

6.2.6 Leaf development

Leaves emerge from groups of founder cells that originate from the peripheral zone of the shoot meristem. The leaf primordium contains cells of all three layers of the meristem. The first sign that a leaf primordium is beginning to form is the disappearance of the KN protein from a cell group still lying in the peripheral zone. The beginnings of the leaf become visible as a lateral hump, caused by periclinal cell division of the L2 layer. If cell division is suppressed, the hump will still emerge regardless. It seems that as the leaf primordium separates from the underlying meristematic tissue the orientation of its cells changes. The newly formed leaf primordium consists of about 100 cells. The total number of cells in a fully grown primary leaf of *Arabidopsis* is estimated to be 130 000. Accordingly, every one of the 100 primordium cells divides about 10 times throughout leaf development.

The leaf is divided into three regions: the blade (*lamina*), the midrib, and the stalk (*petiole*). Each of these has its own different internal structures. In the leaf blade, there is the epidermis, the palisade mesophyll, the spongy mesophyll, and the vascular bundles, which divide it. The order of these tissues reflects the dorsal–ventral axis of the leaf: the dorsal (*adaxial*) side points towards the shoot, whereas the ventral (*abaxial*) side points away. The axis that runs from the base to the tip of the leaf is called *proximodistal*.

Leaf development has been studied extensively by clonal analysis of subepidermal tissues in tobacco. Clones that have been induced within the leaf precursor stretch as wedge-shaped sectors from the base of the leaf along its complete length. Both the upper and lower sides of the leaf can be part of this clone. The borders between sectors are often demarcated by the midrib, but this does not reflect a clonal restriction of cell fate. Moreover, the midrib precursor comprises many cells, whereas the midrib in the finished leaf represents only a small proportion of it. Clones that are induced later in development are limited to proximal sections of the leaf. Later in development, smaller clones have been induced which are scattered all over the leaf blade. At the end of development, no clones could be induced anywhere within the leaf. These data contradict earlier assumptions; according to these, meristems at the leaf margin contribute to leaf development. The truth, however, is that the leaf primordium grows by intercalary cell divisions. At different times, different directions of cell division dominate, and regional rates of cell division vary. In the leaf primordium itself, there seems to be no separate founder cells for the upper and lower sides of the leaf. The leaf margin is much more likely to emerge by interactions of cells within the leaf primordia. Finally, the leaf matures from tip to base: first, the cells at its tip stop dividing, then finally cells at its base also stop. This gradient of development is mirrored in the development of the trichomes (section 6.2.7). Any growth of the leaf now is purely by cell expansion.

Genetic analysis of leaf morphogenesis in *Arabidopsis* has led to the identification of genes involved in directional cell elongation. Mutations in the *ANGUSTIFOLIA* (*AN*) gene result in narrow leaves, although there is no shortage of leaf cells. Rather, the *an* leaf cells do not elongate in the direction of leaf width. Mutations in the *ROTUNDIFOLIA* (*ROT*) genes reduce the length of the leaf, giving it a roundish appearance. In the *rot3* mutant, the number of leaf cells is normal but the cells do not elongate properly along the proximodistal axis of the leaf. The *an rot3* double mutant shows an additive phenotype of leaf morphology. The independent cellular effects of these two genes, *AN* and *ROT3*, suggest that the final size and shape of the leaf depend largely on the geometry of its building blocks, the individual cells, which is subject to many genetic controls.

6.2.7 Development of a cell type in the leaf epidermis: the trichome

The genetics of how a specialized cell type can develop from a uniform cell layer can be studied by investigating trichome development in *Arabidopsis* because, under laboratory conditions a plant can live a perfectly happy and fertile life without its trichomes. Trichomes are differentiated from single cells originating in the epidermis. They occur on stems and sepals, as well as on leaves. However, the leaf trichomes are special because they branch characteristically and are evenly distributed across the leaf's surface.

Trichomes develop from single epidermal cells in young leaf primordia, and have a remarkable cell morphogenesis (Figure **6.9**). First, the trichome cell enlarges, and its DNA content increases until it has eight times more DNA than the surrounding cells. The enlarging cell grows outwardly from the leaf surface, so its nucleus is pushed into a subapical location. When the volume of the trichome cell has increased tenfold, it branches. The nucleus enlarges again

Figure 6.9 Trichome development and mutant phenotypes. Left: the events from the specification of a trichome cell (red) in the leaf epidermis to the mature trichome. To the right are the mutant phenotypes of genes that are important for trichome development. These phenotypes are arranged according to the events (left) that are disrupted in each mutant, respectively. (According to Hülskamp *et al.* 1994.)

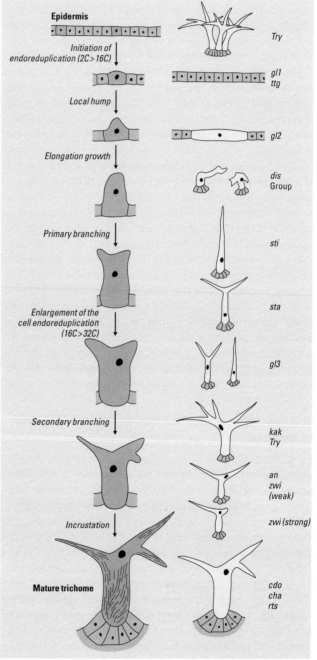

by endoreduplication, while the two newly formed branches grow at their tips. One of the branches elongates to form
the main branch, then branches again. The nucleus then
relocates within the cell so that it lies between the two
branching points. The mature trichome rests on a small
podium formed by a circle of 8–10 epidermal cells, around a
layer of subepidermal cells. The epidermal podium cells differ in shape and size from other cells in the epidermis.

Trichome development is linked to leaf development.
Trichomes emerge from the leaf primordium a distance of
approximately four cells apart. As they grow and branch,
the epidermal cells between them divide. In the end,
neighbouring trichome cells are separated by approximately 30 epidermal cells. The branching of the trichomes is
oriented according to the longitudinal axis of the leaf
primordium. The first-formed branch points towards the
base of the leaf. The later-branching main branch grows
towards the leaf tip. The trichomes of a leaf primordium
are developed to differing degrees: at the leaf tip are trichomes that are already mature, while at the base some trichomes have yet to branch, and in the leaf centre there are
trichomes that have one branch. But how does the characteristic distribution pattern of trichomes on the leaf
emerge?

Two genes that are essential for trichome development,
GLABRA 1 (*GL1*) and *TRANSPARENT TESTA GLABRA*
(*TTG*), have been identified in *Arabidopsis*. The *GL1* gene
codes for a *myb*-like transcription factor, whose mRNA
accumulates in the trichome cells-to-be. The product of the
TTG gene is unknown, but it could be a transcription factor.
An indirect clue to its function is provided by the effect of
the *R* gene in maize, which codes for a transcription factor.
If this gene is constitutively expressed in *Arabidopsis*, more
trichomes emerge even if the *TTG* gene is inactive due to a
mutation. If the constitutively expressed transgene is
expressed as a fusion protein (R–GR) that contains a ligand
binding site of the glucocorticoid receptor (GR), the activity of the R protein can be controlled by a ligand, dexamethasone. In the absence of dexamethasone, the transgenic
ttg plants produce no trichomes. If dexamethasone is
added, every epidermal cell can, in principle, become a trichome. In practice though, the state of development of the
epidermal cell when the dexamethasone is added is important. Young leaf primordia can be induced to produce trichomes all over the leaf blade, whereas completely developed leaves do not respond at all. If, on the other hand,
plants are grown from the start with dexamethasone,

which is later withdrawn, fully developed leaves are covered with trichomes, old leaf primordia are only partially covered, and young leaves have no trichomes at all. The sensitivity of epidermal cells to the R protein correlates with the development of leaves, as it progresses from the distal to the proximal end (section 6.2.6).

Normally the trichomes are not only distributed on the leaf blade in a regular pattern, but there is also only a single trichome at each available site. It is difficult to believe that only a single cell responds to the signal to make trichomes. Indeed, mutations in two genes, *TRIPTYCHON* (*TRY*) and *TTG* can lead to the formation of 'nests' of two or three neighbouring trichomes. If the activity of the *TTG* gene is reduced and the *GL1* gene is expressed constitutively, the effects can be quite impressive. It is believed that when the signal to make trichomes is sent, several neighbouring cells respond to it, but only one of these cells responds successfully, and then suppresses its neighbours. A similar mechanism, called *lateral inhibition*, is employed during the development of single bristles in the epidermis of the fruit fly, *Drosophila*.

If a cell is committed to becoming a trichome, a highly ordered process is set under way. At its conclusion, the characteristic trichome structure will have developed. By mutation of different genes, this cell morphogenesis has been dissected into individual steps (Figure **6.9**, Table **6.3**). The endoreduplication at the start of the process, which leads to the eightfold increase in the DNA content of the diploid cell and its increase in volume, is only dependent on the *GL1* and *TTG* genes that participate in the determination of the trichome cells. A subsequent endoreduplication is triggered by the *GLABRA 3* (*GL3*) gene. Other genes hinder further endoreduplications. The growth of the cell outwardly from the epidermis is promoted by the *GLABRA 2* (*GL2*) gene. This gene codes for a putative transcription factor which contains a homeodomain. In *gl2* mutants, the large trichome cells are often flattened and integrated into the epidermis. This defect can be ameliorated by a mutation that enlarges the trichome cells. The cell size also regulates the onset of branching, a two-step process. An additional product of the gene *STICHEL* (*STI*), is needed in order for the cell to branch. *Sti* trichomes are unbranched, and so, for a given cell volume, are longer than branched trichomes. Other genes are involved in the individual steps during branching. At the end of trichome development, substances that harden the trichomes (*incrustation*) are deposited in the cell.

Gene	Mutant phenotype
GLABRA 1 (GL1)	No trichomes
TRANSPARENT TESTA GLABRA (TTG)	No trichomes
TRYPTICHON (TRY)	'Nests' of trichomes; enlarged trichomes
GLABRA 2 (GL2)	Trichome cells rarely form humps
GLABRA 3 (GL3)	Smaller trichomes
DISTORTED 1 (DIS1)	Trichomes shorter and twisted
DISTORTED 2 (DIS2)	Trichomes shorter and twisted
GNARLED (GRL)	Trichomes shorter and twisted
KLUNKER (KLK)	Trichomes shorter and twisted
SPIRRIG (SPI)	Trichomes shorter and twisted
WURM (WRM)	Trichomes shorter and twisted
CROOKED (CRK)	Trichomes shorter and twisted
ALIEN (ALI)	Trichomes shorter and twisted
STICHEL (STI)	Trichomes long and unbranched
ZWICHEL (ZWI)	Trichomes with only two branches of unequal lengths
ANGUSTIFOLIA (AN)	Trichomes with normal stalk and long branches
STACHEL (STA)	Trichomes with normal stalk and short branches
KAKTUS (KAK)	Enlarged trichomes with several branches
CHABLIS (CHA)	Glass-like trichomes
CHARDONNAY (CDO)	Glass-like trichomes
REISINA (RTS)	Glass-like trichomes

Table 6.3 Genes for trichome development in *Arabidopsis*.

6.3 The generative phase

During normal plant development, flowers emerge from the cells of the vegetative shoot meristem (Figure **6.10**). Initially, the latter produces leaves and lateral shoots more or less continuously, then switches to producing generative organs (such as the flowers). This switch is not only endogenously controlled, there is often also a clear interaction with environmental factors (section 6.3.1). During the process of change induction (*evocation*) the meristem stops growing in length, and is no longer able to produce normal green leaves or side-shoots. With the production of the inflorescences, the plant's vegetative phase has come to an end (Box **6.2**).

The process of flower formation can be split into several discrete steps: the first of these is the induction of flower-

---

Box 6.2 Inflorescence

Depending on the degree of branching, an inflorescence can be either simple (no branching; a single axis), or complex (branched; several axes of the same or higher order). Other differentiating criteria are when and where flowers are made: 'closed' (determinate) inflorescences only produce a single flower at the end of the axis (a terminal flower, as seen in tobacco and petunia); 'open' (indeterminate) inflorescences have no terminal flower at the end of the main axis. In the latter case the terminal inflorescence meristem at the end remains undifferentiated and continuously produces (determinate) flower meristems as well as the primordia for lateral shoots and cauline leaves. Examples of 'open' inflorescences are *Arabidopsis* and *Antirrhinum majus*. In both cases, the molecular genetics of flower development is now well understood (*see also* section 6.3.2).

---

ing, that is the decision to switch from the vegetative to the generative phase. The second step is evocation—transition from the vegetative shoot meristem (VM) to the inflorescence meristem (IM), with the associated generation of the primordia for flowers, lateral organs, and the cauline leaves. The third step is the actual formation of the flower. That is, the transition from inflorescence to flower meristem (FM), with the associated formation of the primordia for the flower organ precursors and their further differentiation. As fourth and last step, the functional phase of the flowering process begins, when the flower organs have gained their characteristic forms and functions. During this functional phase the maturation of the reproductive organs and pollination and fertilization take place. We will

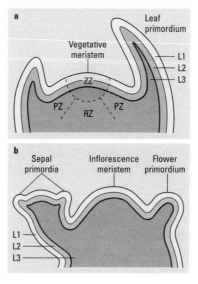

Figure 6.10 Schematic cross-section through the apex of a vegetative shoot (**a**) and flower (**b**). L1, L2, L3, meristematic layers (*see* sections 2.6 and 6.1). PZ, peripheral zone; RZ, file meristem; ZZ, central zone. (According to Huala and Sussex 1993.)

first look at the induction of flowering (section 6.3.1), then at the transitions from VM to IM to FM (steps 2 and 3 of flower formation—section 6.3.2). Step 4, the functional phase, will be discussed in the context of development of the gametophytes (sections 6.3.3 and 6.3.4).

6.3.1 Temperature and photoperiod as inducers of flower formation

Under natural growth conditions, flower formation usually commences when the plant reaches a certain age (it is 'ready' for flowering). The age is, within certain limits, genetically fixed in a species-specific way. Only when this committed stage is reached can flower formation (that is, the change in differentiation from shoot meristem to inflorescence and flower meristem) begin. This is seldom spontaneous; usually it is induced by external factors such as temperature and light (which are very important for flower induction).

Often, flower formation requires that the plant is exposed to low temperatures, usually between a few degrees above zero and 15 °C. The stimulation of flower development by low temperatures is called *vernalization*. Flower development will not necessarily start immediately after the plant has been vernalized, though. Seedlings, or even embryos, can be vernalized; the signal for flower induction is then 'stored' over many cell generations during the entire vegetative stage. The duration of cold, as well as the time at which the low temperature occurs, also influences flowering.

The primary locations where the cold treatment causes an effect are the meristematic tissues. For example, in a handful of cases it has been possible to vernalize isolated embryos growing on artificial media. The plants that grew from these embryos flowered when they emerged. Just cooling the tips of some plants is sufficient to induce flower development. The identity of the participating signal is still unknown. The most straightforward assumption—that the signal for flower induction is transmitted by hormones after vernalization—still awaits rigorous experimental proof.

The second important external factor that can induce flowering is light. In many cases, it is not the intensity, but the duration of the day and night periods (or *photoperiods*) that are important. Depending on the light necessary for flower induction, three main groups of flowering plants can be distinguished: day-neutral (DNP) plants, short-day plants (SDP), and long-day plants (LDP).

The day-neutral plants flower virtually independently of day length. Flower development in these plants is controlled mainly by temperature as well as the age and the nutritional status of the plant. Short-day plants need a certain photoperiod for flower induction, which should fall *below* a certain critical day length. Depending on the latitude, short-day plants often flower in the spring or autumn. Long-day plants, however, only flower when day length rises *above* a critical threshold, so they tend to flower in the summer (Table **6.4**).

The important criterion for inclusion in the short-day or long-day plant groups is not the absolute length of the critical daily photoperiod. Spinach (LDP) and cocklebur (*Xanthium strumarium*) (SDP) flower when day length reaches 14 hours. For spinach, 14 hours is the critical day length, the *minimum* required, whereas it is nearly the *maximum* day length (16 hours) for cocklebur. A longer or shorter photoperiod would have opposite effects on the flowering of these two plants.

It has been shown that plants measure photoperiod through their leaves—a finding that has been interpreted to mean that hormones are involved in signalling the photoperiodic flowering impulse. However, the original model, in which a 'flowering hormone' (called *florigen*) is present in the leaves, which then travels to the shoot tip and induces flower development, could not be substantiated by further experiments.

Further knowledge about the photoperiodic control of flower formation has come from experiments, in which the daily dark period was disturbed by short periods of illumination. Systematic investigations of many SDP and LDP plants have led to the conclusion that it is not the length of the light period but instead the length of the undisturbed dark period that determines whether flowers are formed. The reason for the differing reactions of LDP and SDP plants to the dark-interruption signal is not yet completely understood. Investigations in which the interrupting light

Table 6.4 Examples of obligate short-day plants (SDP) and long-day plants (LDP) with an absolute daylength requirement for flowering.

SDP	LDP
Amaranthus caudatas	Arabidopsis thaliana
Chrysanthemum hort.	Avena sativa
Coffea arabica	Begonia semperflorens
Glycine max	Beta vulgaris
Kalanchoe blossfeldiana	Hordeum vulgare
Perilla ocymoides	Hyoscyamus niger
Setaria viridis	Lactuca sativa
Xanthium strumarium	Vicia faba

was switched on at different times during the dark period have shown that the rate at which plants react to this light is not always the same. It is assumed that the reason for this lies in changes to the plants' activities according to a circadian rhythm, which is regulated by the 'physiological clock' (section 5.4).

The attention of researchers in plant molecular biology has been focused increasingly on the ways in which internal factors participate in flower production, and how they interact with the environmental signals that induce them. Is the endogenous commitment to flowering a general metabolic state, or are very specific genes and gene products involved? Groups investigating this have found clues that suggest an accumulation of specific transcripts during the early phase of flower development. However, without any further functional analyses, it was difficult to decide whether there is a causal relationship with the flowering process. Hints as to the existence and essential roles of certain genes in flowering have been provided by *heterochrony* mutants. These are mutants with precocious or delayed flower induction compared to the wild type. These kinds of mutants have been very well characterized in *Arabidopsis* since their recognition by Rédei (1962). Basically, early and late flowering mutants can be distinguished.

An extreme example of the early flowering group is the recessive mutant *emf* (early maturing flower). Shortly after germination (and having produced very few leaves), this mutant produces a flower. Another flowering-time mutant, *tfl1* (terminal flower) also has a reduced number of rosette leaves at the time when it produces its small inflorescence. Late flowering mutants of *Arabidopsis* have been grouped into two categories according to their response to external factors such as day length and temperature. The first group, with a phenotype largely unaffected by the environment, includes *constans* (*co*), *fd*, *fe*, *ft*, *fwa*, and *gigantea* (*gi*). Some members of the second group are *fea*, *fpa*, *fve*, *fy*, and *luminidependens* (*ld*). They all show delayed flowering, but the flowering time can be shifted earlier under long-day photoperiod and after vernalization. The first molecularly characterized heterochrony gene, *LUMINIDEPENDENS* (*LD*) was isolated using a late flowering mutant made by T-DNA insertion mutagenesis (section 3.2.6). Targeted changes to this gene lead to a retardation in flowering time. This suggests that *LD* participates in determining the time of the onset of flowering. The *LD* gene product has a nuclear localization signal as well as a glutamine-rich region, and so might be a transcription factor (*see* Table **4.2**).

In addition to *LD*, two other late flowering genes have been cloned: *CO* and *FCA*. The derived CO protein reveals two putative zinc finger motifs found in many DNA-binding proteins. The *FCA* gene product contains two RNA-binding domains as well as a putative protein interaction domain (WW domain). Among the genes represented by the early flowering mutants, *TFL1* (as well as its suggested homologue *CEN* (*CENTRORADIALIS*) from *Antirrhinum*) was cloned and found to have similarity to animal phosphatidylethanolamine-binding proteins (PBPs) of thus far unknown function. These few known examples suggest that at least several flowering-time genes may code for transcription factors; others specify products that act at a post-transcriptional level or have a role in as yet to be discovered signal transduction mechanisms.

Among the flowering-time genes that have been characterized so far from *Arabidopsis*, *CO* seems to be the earliest acting. It has a direct effect on two other 'early' floral genes, *LFY* (*LEAFY*) and *TFL1* (*TERMINAL FLOWER*). *LFY*, in turn, is a positive regulator of two subsequent genes, *AP1* (*APETALA 1*) and *AG* (*AGAMOUS*) (section 6.3.2), whereas *TFL1* negatively affects both *LFY* and *AP1*. To further complicate the situation, the early floral gene *CLF* (*CURLED LEAF*) also regulates *ag* expression. It is needed to maintain 'correct' *AG* RNA levels throughout flower development and completely represses the expression in vegetative tissue. Sequence analysis has shown that the derived *CLF* protein has similarity to that of *enhancer of zeste*, a *Drosophila* regulatory gene of the polycomb group. This group of genes has a function in cell fate determination, suggesting that *CLF* fulfils a similar role during flowering.

Flowering-time genes have been found in a number of different plant species. In one particular investigation, a gene was isolated from immature rice flower cDNA. Overexpression of this gene in tobacco plants led to flowering at a time several days earlier compared to control plants. The gene, named *OsMADS1* has a sequence motif, called the MADS box, which also occurs in the regulatory genes of flower organs, and was first recognized there (section 6.3.2).

Flowering-time genes are not just important for the basic understanding of flower formation, but are also of interest in agriculture. We could imagine, for instance, engineering cereals that could be planted more quickly after each other, or planted under less favourable conditions during a short season. On the other hand, the yield of certain varieties could be increased by delaying flowering.

6.3.2 The production of a flower organ—an interaction between homeotic, cadastral, and meristem identity genes

Although the induction of flowering in many plant species is regulated by environmental factors, these are only of limited relevance to the flower formation process itself. Endogenous species-specific developmental programmes take control of flower formation. The roles and importance of essential genes and gene products have been discussed above in relation to flowering-time genes. They can be also shown by a wide variety of mutants which have altered or interrupted flower formation. Two main groups can be distinguished: meristem mutants do not form flowers at all, and organ mutants have altered shapes and numbers of flower organs. As mentioned previously, two model plants have contributed much to molecular investigations of flower formation: *Arabidopsis thaliana* (Brassicaceae), and the snapdragon, *Antirrhinum majus* (Scrophulariaceae). The advantages of *Arabidopsis* have already been discussed (*see* Chapter 3). *Antirrhinum* is also a genetically well-studied species; the transposon mutagenesis technique (*see* section 3.2.7) has been established in this species, and has been used successfully in a number of cases where flower phenotypes were analysed. The large flowers of *Antirrhinum* make crossing experiments easy and facilitate the collection of material for molecular studies. Before looking at the mutant plants of these two model species, we will examine the wild types.

In *Arabidopsis* the leaf primordia are arranged in a spiral, with short internodes at the apex during the vegetative growth phase. This results in its typical shoot rosette (section 6.2.4). After the apical meristem changes into the inflorescence meristem, the spiral leaf primordia arrangement is retained, but a few smaller cauline leaves are produced, and the internodes are much longer. In the axils of the cauline leaves, lateral shoots form. Depending on whether the shoot is first, second, or third order, respectively, a primary, secondary, or tertiary inflorescence is formed. The inflorescence meristem at the tip produces the flower meristems in a spiral arrangement (Figure **6.11**). These develop into a single flower with four concentrically arranged whorls of flower organs. Their primordia are arranged thus, from the outside to the inside (that is, in the meristem from bottom to top): the sepals (whorl 1; calyx), the petals (whorl 2; corolla), the stamens (whorl 3; androecium), and the carpels (whorl 4; gynoecium).

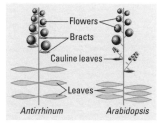

Figure 6.11 Schematic drawing of *Antirrhinum* and *Arabidopsis*. The typical organization of above-ground organ systems of fully grown flowering plants is shown. To give a better view, the lateral shoots in the leaf axils have not been drawn. The positioning of the leaves (spirally arranged in *Arabidopsis*; pairs of opposite leaves at each node in a decussate arrangement and spiral bracts in *Antirrhinum*) is not shown in this two-dimensional illustration. (Taken from Coen, E. S. and Meyerowitz, E. M. *Nature*, **353**, (1991), 31–7.)

Whorls 1 and 2 are the sterile parts of the perianth (the flower's outer covering), while whorls 3 and 4 represent the flower's reproductive organs. The precise arrangement of the four sepals, four petals, six (four short and two long) stamens, and two carpels that fuse to form the ovary can be seen in Figure **6.12**.

Antirrhinum has pairs of opposite leaves at each node. Each pair is arranged at a right angle to the previous one (decussate phyllotaxis) with long internodes. After the vegetative meristem has changed to the inflorescence meristem, very small cauline leaves (bracts) are produced in a spiral arrangement with short internodes (Figure **6.11**). In the axils of the cauline leaves, a flower meristem is established, which produces the flower organs arranged in whorls (Figure **6.12**). However, the number, shapes, and sizes of these differ from those of *Arabidopsis*. *Antirrhinum* has five sepals and petals, four stamens, and two carpels which fuse. The bottoms of the petals fuse (sympetalous corolla) and form the *flower tube*, whereas the apical regions (lobes) remain unfused. The shapes of the two upper lobes differ from those of the three lower ones. In the connecting area between the upper and lower lobes there is a kind of hinge that enables the flower to open to admit insects (section 6.3.4). The third whorl contains five stamen primordia, the uppermost of which degenerates prematurely. Of the four remaining stamens, the upper two are shorter than the two lower ones.

The flower of *Antirrhinum* is *zygomorphic*. It has a vertical axis with bilateral symmetry. The shape of the flower with its special changes to the corolla is an adaptation to insect pollination. The *Arabidopsis* flower is wind pollinated and, with the exception of whorl 3 (with its four plus two stamens), has a number of axes of symmetry that pass through the centre and divide the flower longitudinally

Figure 6.12 Floral diagrams of *Antirrhinum* and *Arabidopsis*. The four flower whorls contain, from the outside to the inside: sepals (whorl 1), petals (whorl 2), stamens (whorl 3), and the carpels that have fused to form the ovary (whorl 4). The small circles within the ovary indicate the ovules (*see* section 6.3). (From Coen, E. S. and Meyerowitz, E. M. *Nature*, **353**, (1991), 31–7.)

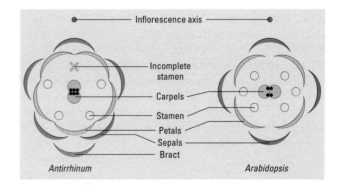

Inflorescence axis

Incomplete stamen

Carpels

Stamen

Petals

Sepals

Bract

Antirrhinum

Arabidopsis

into two halves (*actinomorphic flower*). However, if whorl 3 is taken into account, then there are only two axes of mirror-image symmetry (Figure **6.13**). The flower of *Antirrhinum* is about 10 times the size, and 1000 times the weight, of the *Arabidopsis* flower.

A number of flower organ identity mutants (floral homeotic mutants; **Box 6.3**) have been found so far in *Arabidopsis* and *Antirrhinum*. In most of them, two neighbouring whorls are affected by the mutation, which led to the idea that there are three classes of genes that confer flower identity functions (A, B, and C). Function A

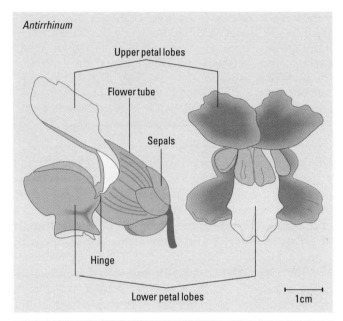

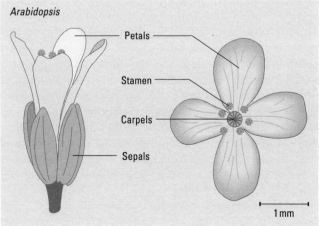

Figure 6.13 Flowers of *Antirrhinum* (top) and *Arabidopsis* (bottom), each viewed from the side (left) and from the front/top looking towards the flower base (right). The *Antirrhinum* flower (top left) is drawn slightly opened to show the function of the hinge between the top and the bottom petals. The *Arabidopsis* flower is turned by 45° relative to the diagram of the flower in Figure 6.12. Note the different sizes as indicated by the scale bars on the right. (According to Weberling 1989.)

Box 6.3 Homeotic flower mutants

Flowers are characteristic and taxonomic-
ally reliable features of higher plants.
Therefore, changes in flower morphology
tend to be easily perceived. Descriptions of
abnormal flowers with altered numbers and
arrangements of single flower organs can be
found in ancient literature, more than 2000
years old. The idea that such altered flower
structures may be important tools for the
understanding of normal flower development was put forward in the mid-eigh-
teenth century by Linnaeus. Goethe (in
1790) used the term *abnormal metamorphosis*
in conjunction with flower development.
He used this term to describe the situation
whereby out-of-place flower organs are
formed as substitutes for the normal flower

organs. Goethe presumed that all four
flower organs are just modified leaves. In
his opinion, the specific differentiation was
affected by a special substance rising up
from the plant below into the developing
flower. The term *homeosis* was used by
Bateson in 1894 to describe the phenomenon of organ identity (that is, the formation
of the right organ in the right place) in his
descriptions of animal systems. A *homeotic
mutant* is an individual with a disrupted
organ identity. *Homeotic genes* are therefore
those genes that lead to (via their gene
products) the formation of the 'correct'
organs in the right place. This terminology
has become widely accepted and is today a
general term used in developmental biology.

---

affects the outer area representing whorls 1 (sepals) and
2 (petals), function B affects the middle area representing whorls 2 and 3 (stamens); and function C affects the
inner area representing whorls 3 and 4 (carpels). A loss
of function A results in the formation of carpels instead
of sepals in whorl 1 and stamens instead of petals in
whorl 2. This is the case in the *Arabidopsis* mutants *apetala 1* (*ap1*) and *apetala 2* (*ap2*), and in the *Antirrhinum*
mutant *squamosa* (*squa*). A lack of function B results in
the substitution of sepals for petals in whorl 2 and
carpels for stamens in whorl 3. The *Arabidopsis* mutants
apetala 3 (*ap3*) and *pistillata* (*pi*) and the *Antirrhinum*
mutants *deficiens* (*def*) and *globosa* (*glo*) belong to this
class. If function C is missing, the third whorl forms
petals instead of stamens and the fourth whorl produces
sepals instead of carpels. This is found in the *Arabidopsis*
mutant *agamous* (*ag*) and in the *Antirrhinum* mutant *plena*
(*ple*) (Table **6.5**). Genetic analysis of double mutants has
hinted that the activity of genes from class B is independent of genes from classes A or C. This does not hold
true, though, for genes from classes A and C. The phenotypes of mutants lacking function C show an enhanced
function A, and vice versa.

   Arabidopsis mutants have been made with defects in all
three (A, B, and C) functions. Such mutants (genotype: *ap2*

Table 6.5 Phenotypes of some organ identity mutants from *Arabidopsis* and *Antirrhinum*.

	Phenotype of the flower whorl			
Genotype	1	2	3	4
Wildtype	sepals	petals	stamens	carpels
ap2, ovu	carpels	stamens	stamens	carpels
ap3, pi, def, glo, sep	sepals	sepals	carpels	carpels
ag, (ple)	sepals	petals	petals	sepals

Arabidopsis mutants: *agamous* (ag), *apetala 2* (ap2), *apetala 3* (ap3), *pistillata* (pi). *Antirrhinum* mutants: *deficiens* (def), *globosa* (glo), *ovulata* (ov), *plena* (ple), *sepaloidea* (sep).
The organ phenotype of whorl 4 of the *plena* mutant can be similar to sepals, petals, or carpels.
ag and *ple* form more perianth-like whorls inside whorl 4.

pi ag) still give rise to something that resembles a flower from its outer appearance, but there is no flower organ identity. All flower organs are similar to normal vegetative leaves—they are green and are covered with trichomes (section 6.2.7). We can conclude that this is the ground state of flower organs in the absence of influence from their controlling genes. These and other investigations of flower *organ identity* mutants have led to the formulation of rules that are the basis for the *ABC model* (Meyerowitz and colleagues; Figure **6.14**). The model explains the vast majority of phenotypes of homeotic flower mutants, using only three axioms:

1. Each whorl of a wild-type flower is specified by the combination of functions A, B, and C. At least one, but at most two, of these functions are sufficient if they are actively expressed. Function A activity alone leads to sepal production, functions A and B together lead to petals, B and C to stamens, and C alone to carpels. A double AB-defective mutant (for example, *ap1 ap3*) of *Arabidopsis* has only C function activity in all four whorls, so only carpels will be produced.

2. The effect of each individual organ identity function is independent of the affected flower region. Activity of function A alone causes sepal formation in whorl 1 of the wild type. The *Arabidopsis* mutant *ap3* lacks activity of function B, so whorl 2 has only function A activity, resulting in the formation of sepals instead of petals, as in the wild-type whorl 1.

3. A and C activity inhibit each other: mutants lacking one function show an increase in the other function. Examples are mutants with defective C function, for instance *ag* in *Arabidopsis* or *ple* in *Antirrhinum*. In these, petals are not only formed in whorl 2, as in the wild type, but also in whorl 3. Whorls 1 and 4 form sepals.

Figure 6.14 The simplified ABC model of the flower organ identity genes. The model has been developed in this form for *Arabidopsis* and in a slightly different way for *Antirrhinum* The scheme shows an arbitrary longitudinal section through the flower and represents all four types of flower organs: sepals in the outer whorl, petals in the second whorl, stamens in the third whorl, and carpels in the inner whorl. In the wild type, the genetic A, B, and C functions are locally distributed such that A is effective in whorls 1 and 2, B is effective in whorls 2 and 3, and C is effective in whorls 3 and 4. A alone forms sepals, A plus B form petals, B plus C forms stamens, and C alone forms carpels. The A and C functions coexist in a kind of equilibrium, counteracting each other. If one is missing, the function of the other will be extended. There is no inhibitory effect between the B and A/C functions. In mutants that lack A function (a⁻) the extension of the C function to whorls 1 and 2 leads to the formation of carpels and stamens. The b⁻ genotype that lacks B function only forms flower organs that are determined by the A or C functions: sepals in whorls 1 and 2 (A function) and carpels in whorls 3 and 4 (C function). The c⁻ genotype has, because of the extension of A function, petals and sepals in whorls 3 and 4, respectively. Investigations with mutants and transgenic plants have confirmed the validity of this model. However, some further assumptions are required to explain all experimentally observed results (see text). (According to Weigel and Meyerowitz 1994.)

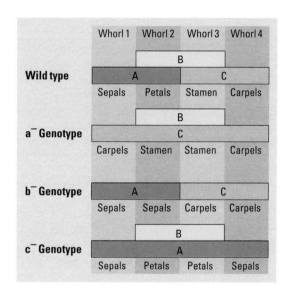

Since studies with mutants have shown that the regulatory genes mentioned above are essential for the identity of flower organs, the corresponding wild-type genes have been isolated and characterized (*see* Chapters 3 and 4). The first of these genes to be cloned were *DEFICIENS* (*DEF*) from *Antirrhinum* and *AGAMOUS* (*AG*) from *Arabidopsis*. Since then, most, if not all, known organ identity genes from these two plant species have been molecularly characterized. The equivalent pairs of genes (orthologs) from the two species are (in the order *Arabidopsis/Antirrhinum*): *AGAMOUS* (*AG*)/*PLENA* (*PLE*), *APETALA 3* (*AP3*)/*DEFICIENS* (*DEF*), *PISTILLATA* (*PI*)/*GLOBOSA* (*GLO*), AND *APETALA 1* (*AP1*)/*SQUAMOSA* (*SQUA*). Only *APETALA 2* from *Arabidopsis* has no *Antirrhinum* counterpart that has been sequenced so far (Table **6.6**). These gene pairs all code for proteins having certain characteristics in common.

A conserved region from the deduced amino acid sequences shows local similarity to a part of the keratin protein of animals, and so has been called the K domain. Because of features in its sequence, this region is able to form an amphipathic helix (coiled coil) and may enable protein–protein interactions. Moreover, nearly all the deduced proteins from the flower genes mentioned above have a conserved region (called a MADS box) close to their amino-terminal end (*see* Table **4.2**). This is responsible for the protein's DNA binding and dimerization capabilities (Figure **6.15**).

	Arabidopsis	*Antirrhinum*
Organ identity genes		
A function	AP1	(SQUA)
	AP2	?
B function	AP3	DEF
	PI	GLO
C function	AG	PLE
Cadastral genes		
	AG	PLE
	AP2	?
	SUP	?
	LUG	?
	UFO	FIM
Meristem identity genes		
	UFO	FIM
	AP1	SQUA
	AP2	?
	CAL	?
	LFY	FLO
Flowering-time genes		
	EMF	?
	LD	?
	TFL1	?

Table 6.6 An (incomplete) list of genes that control flower development.

Note that several genes can be assigned to more than one functional group. Some genes seem to be unique to one species (e.g. *AP2*). Others reveal species-specific differences. For instance, *AP1* and *SQUA* both act as meristem identity genes, whereas the function of *AP1* as an organ identity gene is less evident, if at all, for *SQUA*.

The sequence similarities of flower proteins with animal and fungal transcription factors allow us to draw certain conclusions about their functions. Yeast *MCM1* and mammalian *SRF* (*see* Table **4.2**) interact with similar binding areas of certain promoters. They both contact (or bind to) a DNA sequence motif, $CC(A/T)_6TGG$ (the CArG box). The plant MADS box factors also bind to this sequence, as was shown in *in vitro* experiments with the AG protein. DNA binding activity is not a feature of the monomeric protein—the MADS box proteins exist as homodimers (such as AG—see above) or heterodimers. On their own, the DEF and GLO proteins of *Antirrhinum* do not possess the ability to bind DNA specifically; they can only do this when they work together in concert. Such heterodimerization as a prerequisite for DNA binding has also been shown in the case of the AP3 and PI proteins from *Arabidopsis*. It is assumed that the different MADS box proteins interact via the K domain (see above). Moreover, further interactions with

SRF	[141]R GRVK IKMEF IDNKLRRYT TFSKRKTGIMKKAYEL STLTGTQVL LLL VASETGHVYTFATRKL[202]
MCM1	[16]KERRK IEIKF IENKTRRHVTFFKRKHGIMKKAF EL SVLTGTQVL LLL VVSETGL VYTFSTPKF[77]
DEFA	[1]MARGK IQIKR IENQTNRQVTYSKRRNGL FKKAHEL SVLCDAKVS IIMISSTQKLHEYISPTT[62]
GLO	[1]MGRGK IEIKR IENSSNRQVTYSKRRNGIMKKAKEI SVLCDAHVSVIIFASSGKMHEFCSP ST[62]
SQUA	[1]MGRGKVQLKR IENKINRQVTFSKRRGGLLKKAHEL SVLCDAEVAL IVFSNKGKLFEYSTDSC[62]
PLE	[14]NGRGK IEIKR IENITNRQVTFCKRRNGLLKKAYEL SVLCDAEVALVVFSSRGRLYEYANNSV[75]
AG	[51]S GRGK IEIKR IENTTNRQVTFCKRRNGLLKKAYEL SVLCDAEVAL IVFSSRGRLYEYSNNSV[112]

Figure 6.15 Amino acid sequence at the amino-terminal end of MADS box proteins. The amino acids shaded in red are conserved in at least 50 per cent of these proteins. Proteins of this type probably act as transcription factors, and the MADS box domain is required for DNA binding and protein dimerization. (According to Schwarz-Sommer et al. 1992.)

regulatory proteins can happen, leading to positive or negative regulation of the expression of the target gene. An example of this type of combined regulatory interaction is provided by MCM1. This regulatory protein occurs in haploid yeast cells of both mating type a (MATa) and mating type α(MATα) but, depending on the cell type, its action differs. In MATa cells, MCM1 alone switches on a-specific genes. In MATα cells, however, it binds another protein (MATα1), and together they activate promoters of MATα-specific genes. At the same time, MATa-specific genes are repressed by a MCM1–MATα2 complex.

These in vitro findings suggest that all flower MADS box genes investigated to date act in vivo as sequence-specific DNA-binding proteins and transcription regulators. However, this does not mean that all factors that control flower development are MADS box proteins. For example, the AP2 protein of Arabidopsis does not contact a MADS box. Instead, it contains two copies of a 68-amino-acid motif, designated the AP2 domain, which is conserved in a number of other plant genes. Its core region is capable of forming an amphipathic α-helix. The AP2 domain is related to the DNA-binding region of ethylene response element binding proteins (EREBPs) (see section 5.5).

Most flower identity genes are expressed in a locally defined area, as can be shown by in vitro hybridization (see section 3.3.2). This area is equivalent to the area of the flower that, when there is a defect in its gene, is homeotically changed. Gene expression analysis thus confirms the genetic data and supports the ABC model. For example, it was shown that lack of function B initially has no visible effect on the expression area of genes for the A or C functions. However, genes for the A and C functions influence each other's patterns of expression: the Arabidopsis gene AP1 (for function A) is more strongly expressed in whorls 3 and 4 of ag mutants (lacking function C). On the other hand, an ap2 mutant showed stronger expression of *AG* (for functions A and C) at the RNA level in whorls 1 and 2.

Although these results are in agreement with the ABC model, other investigations on expression of these genes do not concur. Transcripts of the *AG* gene have been found in wild-type *Arabidopsis* plants as expected in whorls 3 and 4, and in mutants with a defect in function A in all four whorls. What was unexpected was the presence of *AG* transcripts in whorls 3 and 4 in the *ag* mutant *ag1*. The *AG* expression pattern does not correlate with the genetically defined C function that the mutant lacks. Another atypical example is *AP2* (for function A). *AP2* RNA could be shown to be present in all four whorls of wild-type *Arabidopsis*. However, genetic A function is only seen in whorls 1 and 2. These and other findings have led to the conclusion that additional control genes must be present, whose products act locally to repress or enhance functions.

The control mechanisms for the establishment the identities of flower organs are currently under intense investigation. One current question concerns the comparison between protein and RNA levels. As we have seen in Chapter 4, the transcripts of a particular gene can be present in similar amounts in different organs, but can lead to production of different amounts of protein. A protein's function can be strongly affected by post-translational modifications, such as phosphorylation. Clones of other flower control genes from the model plants *Arabidopsis* and *Antirrhinum*, as well as from other species (petunia, oilseed rape, maize, and tobacco) have also been analysed. Together, the results from these two lines of investigation should soon lead to a further refinement of the ABC model.

The study of overexpression of individual genes in transgenic plants has been especially helpful. Using this technique, homeotic changes to flower organs have been made. These 'gain of function' experiments are necessary as positive proofs of gene functions. They complement the loss of function experiments (the genetic analyses of mutant plants) and reinforce their conclusions. Also, this approach makes it possible to characterize further control genes for which there are no known mutants. In combination with the use of mutants, transgenic plants also open up practical applied approaches. For *Arabidopsis* there is a 'kit' already available consisting of mutants that lack the A, B, or C functions, as well as transgenic plants with an enhanced B or C function. This kit allows, in principle, the creation of 'desired' floral organs in any of the four whorls. The crossing of the resulting mutant plants leads to predictable and novel flower structures in the progeny.

Homeotic mutants with altered organ identities are, at the same time, pattern mutants (*see* section 2.5.2). They share these features with the so-called meristic mutations which lead to altered numbers of structures and organs (for example, the *clavata* mutants of *Arabidopsis* that affect the *CLV1* or *CLV2* gene loci). It is not always easy to distinguish between these classes of mutants. For example, some homeotic mutations lead to altered numbers of flower organs. In a third class of mutations, both the number and identity of the flower organs remain unchanged, but organ development stops prematurely. Mutants with organ development interrupted in this way can also be pattern mutants. Still another group of mutants is involved in the control of floral symmetry. Two well-known examples are the *cycloidea* (*cyc*) and *radialis* (*rad*) mutants from *Antirrhinum*. *Cyc* mutants have a (semi-)peloric phenotype, that is the resulting flower reveals radial rather than dorsoventral symmetry. Except for a nuclear localization signal, the derived protein has no homology with any other sequenced gene product. *Cyc* is expressed very early in the dorsal region of the floral meristem.

Unlike *cyc*, most mutants discussed so far have something in common: their phenotype is established relatively late in flower development. The question is whether expression of the corresponding wild-type genes is regulated in time and space by earlier-acting developmental genes. The so-called cadastral genes (whorl-identity genes) belong to this earlier class. Their gene products are regulators that act locally. Because of the expression patterns of such regulatory genes in the wild-type and mutant plants, it is assumed that they limit the expression of the later homeotic organ identity genes to specific areas of the flower and keep them out of other areas. Genes with cadastral function are *SUPERMAN* (*SUP*) and *LEUNIG* (*LUG*) from *Arabidopsis*, and *PLENA* (*PLE*) from *Antirrhinum*. For example, the *SUP* gene product inhibits the organ identity gene *PI* (Table **6.6**).

The flower meristem identity genes have an even earlier function than the cadastral genes. These genes are positive inducers of the organ identity genes. In mutants where these genes are inactivated, there is either only a partial or a complete lack of flower formation. In extreme cases, such mutants form shoots instead of flowers. Typical flower meristem identity genes are *LEAFY* (*LFY*) and *CAULIFLOWER* (*CAL*) in *Arabidopsis*, and *FLORICAULA* (*FLO*) and *SQUAMOSA (SQUA)* in *Antirrhinum* (Table **6.6**).

Moving further down the hierarchy of flower control genes, we come across the inflorescence meristem and, probably directly, the flowering-time or heterochrony genes (section 6.3.1). During flower formation, we have therefore a chain of control genes which influence, via their gene products, the genes at the next level in the cascade. Genes at a particular level act in combination so to enhance or suppress each other's effects. However, this portrait of the genetic control of flower development is still incomplete. It is already known that many of the participating genes act at more than one level. For example, *AP1* in *Arabidopsis* is, on the one hand, an organ identity gene that specifies the A function, and, on the other hand, it acts as a flower meristem identity gene. This latter function seems to predominate in the case of *SQUA*, the *AP1* ortholog in *Antirrhinum*. *AG* (*PLE* in *Antirrhinum*) has a dual role as an organ identity gene (function C) and a cadastral gene; and, finally, *AP2* has three functions: as an organ identity gene (function A), as a cadastral gene, and as a flower meristem identity gene (Table **6.6**). Similarly, genes that are not involved in floral organ identity but act earlier can have multiple functions. For instance, *UFO* (*UNUSUAL FLORAL ORGANS*) in *Arabidopsis* (*FIM* (*FIMBRIATA*) in *Antirrhinum*) is a cadastral gene that mediates between meristem and organ identity genes. In addition, it is involved in establishing the whorled pattern of floral organs and controlling the determinate growth of the floral meristem.

The network of flower control genes is reminiscent of cascades of regulatory genes in animal systems. Perhaps the best-known example of these comes from the development of *Drosophila* larvae, in which the chain of regulatory interactions has been extremely well investigated. The earliest participating genes are maternal control genes. These activate the first segmentation genes (*gap*), which activate in turn the pair-rule genes and segment polarity genes. Only relatively late in development are the homeotic selector genes activated, turning on the effector genes that are responsible for all the important cell functions.

A comparison of the flower organ identity genes with the homeotic selector genes of *Drosophila* shows that there are similarities as well as differences. Both groups of genes code for transcription factors. In the case of *Drosophila*, they are the *homeobox proteins*, which have a conserved homeotic domain. In the flower proteins, we find mainly MADS box motifs (see above). The selector genes of animals are usually grouped together with a common control region in the genome, whereas all flower organ identity genes investigated to date exist as single genes in the genome. What the two groups of animal and plant homeotic genes have in common is that apparently they need to be active over a certain length of time, not just transiently. In the network of developmental control genes, both groups are rather late genes. Early control genes in animal systems act transiently for the most part. As we have seen, the cadastral genes that are involved in controlling the later flower organ identity genes (see above) form such a group of early flower control genes. They have formal similarities to the *gap* genes of *Drosophila* which control the homeotic selector genes.

To date it is unknown whether the products of the late flower organ identity genes act directly on the effector genes for central cell function, or whether additional steps are involved. The different types of flower organs vary in their gene expression patterns at the RNA or protein levels. This can be seen, for example, in quantitative DNA/RNA hybridization and two-dimensional protein gel electrophoresis (see section 2.4 and Chapter 4). Even in an individual organ system, depending on the stage of development, different expression patterns have been discovered (see, for example, the stamens, section 6.3.3).

6.3.3 Production and development of the male and female gametophytes

The previous section described the events from floral induction to the formation of the flower organs. These events are accompanied by drastic morphological and physiological changes. However, once the functional flower has formed there are no major changes that could be readily seen from outside. All the important steps in this functional phase of flower development happen internally. As we have seen in section 2.2.3, the most important step in this phase is the production of meiospores—the extremely reduced forms of micro- and megagametophytes. Pollination and fertilization are two more steps in this phase which lead to the formation of the zygote and later to embryogenesis. Embryogenesis (the initial phase of sporophyte formation, *see* section 2.2.3) has already been discussed in section 6.1. Now, we will focus on the preceding reproduction events in the flowers of angiosperms (sections 6.3.3 and 6.3.4). It has proven helpful to separate the entire process of flower formation into several single steps by using features that can be easily recognized from out-

Phase	Typical features at the beginning of the phase	Duration (hours)	Age at end of phase (days)
1	Flower bud emerges	24	1
2	Flower primordium forms	30	2.25
3	Sepal primordium emerges	18	3
4	Sepals cover flower meristem	18	3.75
5	Petal and stamen primordia visible	6	4
6	Sepals cover the flower bud	30	5.25
7	Primordia of the long stamens form stems at their bases	24	6.25
8	Pollen sacs on long stamens	34	7.25
9	Petal primordia form stems at their bases	60	9.75
10	Petals have the same heights as short stamens	12	10.25
11	Stigma papils become visible	30	11.5
12	Petals have the same height as long stamens	42	13.25
13	Bud opens, petals become visible, flower opens (anthesis)	6	13–14
14	Long anthers higher than the stigma	18	15
15	Stigma higher than long anthers	24	16
16	Petals and sepals start withering	12	16.5
17	All flower organs drop off and the silique is green	192	24.5
18	Silique turns yellow	36	26
19	Silique starts to open	24	27–28
20	Seeds drop out		

Table 6.7 Typical features that can be seen from outside during flower formation in *Arabidopsis*.[a]

[a] These are only approximate values, the duration depends on the ecotype used and the various external factors involved, such as temperature and photoperiod.

side (Table **6.7**). Detailed investigations of the fertile flower organs have been carried out at these various steps. Therefore, using this stepped scale of development, it is possible to draw precise conclusions about the internal reproduction events within the flower at each stage.

Male gametophyte

The stamens of whorl 3 each consist of a filament and an anther at their upper end. The latter contains a sterile central region (the *connective*) that supports two thecae. Each theca is composed of two pollen sacs. The stamens of angiosperms are equivalent to an entire microsporophyll of a heterospore fern. The pollen sac is analogous to the microsporangium. Dissection of the pollen sac shows that its wall is composed of four layers. These are, from the outermost to the innermost: the epidermis, the endothecium, the middle wall layer, and the tapetum. They surround the pollen-forming archespor.

Molecular analysis of anther development hinted that there is a class of genes—called *TA* genes—whose products are specific for the anthers (that is, their products exist mostly, if not exclusively, within the anthers). Some of these genes code for lipid transferase proteins of the tapetum (*TA32* and *TA36*). Another, *TA56*, codes for a thiolendopeptidase in a limited area of the connective. Although not all genes belonging to this group were characterized by their function, they have unique expression patterns in space and time.

The pollen mother cells of the archespor divide meiotically (microsporogenesis). Each produces four haploid pollen cells (or *microspores*; Figure **6.16**). The protoplast of these cells is surrounded by a resistant two-layered wall called the *sporoderm*. The inner layer (the *intine*) contains the typical components of plant cell walls. However, the outer layer (the *exine*) consists of sporopollenins, which are highly resistant to chemical attack.

The development of the extremely reduced male gametophyte starts with an unequal mitotic division of a pollen

Figure 6.16 Formation of spores and gametes. Pollination and fertilization in the reproductive organ system of the flower. (According to Drews and Goldberg 1989.)

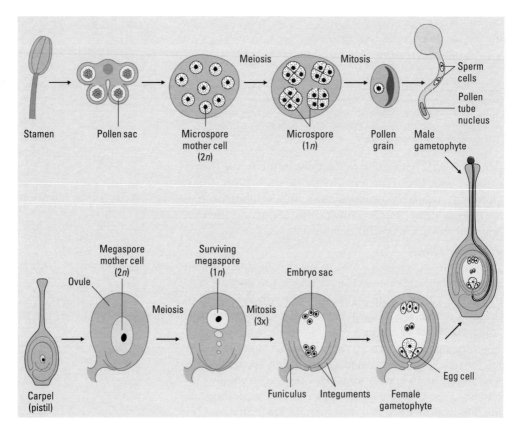

cell with a single nucleus (Figure **6.16**). This leads to a larger vegetative cell and a smaller generative cell. In many plant species the latter usually lacks plastids. This explains why the plastome is inherited maternally in these species (*see* section 4.2.1). The generative cell initially adheres to the wall of the vegetative cell. Later it loosens, becomes spiral, and is surrounded by the vegetative cell, which by now nearly fills the entire pollen grain. By a further mitotic division, the generative cell produces two sperm cells whose cell nuclei are surrounded by only a thin layer of cytoplasm. Although the first mitotic division usually happens within the pollen sac, this does not hold true for the second division. After the pollen sac opens, the pollen grains are released (dehiscence). Under the right conditions they 'germinate' on the stigma of the pistil (section 6.3.4, Pollination). The vegetative cell grows out of the exine and forms the pollen tube. In many cases, only now does the division of the generative cell begin. The germinated pollen grain is the end product of the male gametophyte's development. It can be viewed as a highly reduced equivalent of the fern microprothallium.

For the genetic analysis of the male gametophyte, a large number of male-sterile nuclear mutants are available (cf. mitochondrially transmitted pollen sterility; *see* section 4.3.1). Within this group of mutants (which have a defect in the development of stamens but not of carpels), a further distinction can be made according to the type of defect:

(1) homeotic mutants such as the *pi* and *ap3* mutants from *Arabidopsis* discussed in section 6.3.2;
(2) mutants having a defect in microsporogenesis, for instance the *Arabidopsis ms* mutants; and
(3) mutants with a defect either in pollen release (*Arabidopsis msH*) or pollen function (*Arabidopsis pop*).

The majority of these mutants are sporophytic, which means *MS/ms* heterozygotes produce normal pollen. Only in a few cases have gametophytic mutants been obtained whose heterozygotes produce only 50 per cent functional pollen.

Female gametophyte

In angiosperms the carpels of the fourth flower whorl fuse at their bases to form the ovary. The upper areas of the neighbouring carpels can fuse to varying extents, depending on the species. Often they are completely joined and the resulting structure is then called a pistil. On top of the basal ovary are the central style and the apical stigma.

Depending on species, one or more ovules (seed progenitors) form inside the ovary. Each ovule is connected to the placenta by the *funiculus*, which contains vascular tissue and helps to nourish the ovule. The outer coat of the ovule is built from two integuments, the inner of which is called the *endothelium*. They start at an area at the base of the ovule called the *chalaza*, which marks the border with the funiculus and which surrounds the inner tissue area (*nucellus*) of the ovary. Only a small spot of tissue opposite the chalaza is not covered by the integuments, leaving an opening called the *micropyle*. The carpels are analogous to the macrosporophylls of the ferns, and the nucellus is similar to the macrosporangium.

In the nucellus, a single cytoplasm-rich cell enlarges—the embryo sac mother cell. It divides meiotically (microsporogenesis; Figure **6.16**). Of the four emerging haploid spores, only the one that lies in the direction of the chalaza will lead to the female gametophyte. This cell with its single nucleus (the *megaspore*) grows to form the embryo sac. Then, in typical cases, megagametogenesis begins with three consecutive nuclear divisions. The products of the first division (of the primary embryo sac nucleus) move to opposite poles of the embryo sac; once there they divide twice again. There are now eight nuclei in total. Three of the four nuclei at each cell pole remain there, and thin walls are formed which separate them (together with some cytoplasm) from the rest of the embryo sac. The three upper cells nearest the micropyle form the egg apparatus, which consists of the large egg cell and the two smaller synergids. The three lower cells are antipodals. The two remarking nuclei (polar nuclei) migrate to the centre of the embryo sac and once there fuse together to form the secondary diploid embryo sac nucleus. In some plants this is surrounded by cytoplasm and a wall, to form the *central cell*. Altogether, the mature embryo sac is the end product of female gametophyte development; it is equivalent to the fern megaprothallium.

Molecular analysis of the female gametophyte has long been hampered, mainly because the material is difficult to get hold of in sufficient quantity. Differential mRNA display is a new technique to examine mRNA populations using reverse transcription–polymerase chain reaction (RT-PCR) amplification. Because it needs little starting material, it should be an ideal tool for analysis of the female gametophyte and so should help in understanding the role of cell-specific gene products. Some important hints in this direction have come from flower mutants that lack specific carpel functions and whose female parts are usually sterile. A number of *Arabidopsis* mutants with defects in development of the integuments, the ovule, and/or embryo sac all belong to this class. The examples that have been studied in detail include the mutants *short integuments -1 (sin1)*, *ovule mutant-2* or *-3 (ovm2, ovm3)*, *bell (bel1)*, and *aberrant testa shape (ats)*, as well as two ovule mutants designated 47H4 and 54D12. The *BEL1* gene was cloned and found to encode a homeodomain protein localized to the nucleus. Two mutants that were mentioned earlier (section 6.3.2), *ap2* and *sup*, not only affect the general floral architecture but also interfere with ovule development. Recently, another ovule mutant from *Arabidopsis*, designated *aintegumenta (ant)*, was described and the *ANT* gene cloned. The gene product shares high sequence similarity with AP2 in the so-called AP2 domains, suggesting that these two proteins are related transcription factors. The above mutants are all sporophytic, but their gametophyte development is also affected. Gametophytic mutations are rare, examples are the maize mutant *indeterminate gametophyte (ig)* and the *Arabidopsis* mutant *Gf*.

6.3.4 Pollination, fertilization, and formation of the zygote

Pollination

In gymnosperms the pollen directly contacts the micropyle of the ovule. The closed ovaries of angiosperms make this impossible: pollen sticks to the papillar cells on the surface of the stigma. Cell–cell interactions follow and compatible pollen grains germinate and form a pollen tube. The pollen tubes grow through the papillar cell walls and, via the middle lamellae of the pistil, reach the micropyle of the ovule. The molecular and cellular reasons for this highly targeted growth of the pollen tube are still largely unknown.

The pollen grains germinate on the stigma, a process that is under strict genetic control so that only pollen of a similar or closely related species is allowed to germinate. Many plants also possess self-incompatibility (SI) mechanisms that hinder self-pollination (in other words, they are self-sterile). These mechanisms are linked to the activity of the *S* genes which are localized at the S locus. The SI mechanisms can differ greatly, depending on species. Depending on whether the genotype of the haploid pollen or the diploid pollen donor determines the success of the pollination, it is termed *gametophytic SI* or *sporophytic SI*. Many crucifers are of the sporophytic SI type. For example, for *Brassica napus* (oilseed rape) it has been shown that the S locus has genes for two proteins for the stigma surface that are functionally related—one is a secreted glycoprotein (SLG; S locus glycoprotein) and the other a receptor kinase (SRK; S locus receptor kinase). The SI mechanism of this system is probably as follows: under the influence of a pollen factor (either a protein or a low molecular weight ligand), the activity of the SLG protein is changed so that it can interact specifically with the SRK component. This means that the subsequent chain of events is interrupted; ordinarily, this chain would lead to the successful formation of a pollen tube. In Solanaceae, which typically possess a gametophytic SI, the S locus codes for a glycoprotein with ribonuclease activity. Clearly, the Solanaceae use a different molecular mechanism than do the Cruciferae.

The flowers of most angiosperms contain stamens as well as carpels, and are thus optimally equipped for pollination. However, sometimes both types of flower organs are formed initially, but the development of one or the other is subsequently interrupted (sex determination). The example of the monoecious plant *Zea mays* showed that the determination of a flower's sex as male or female is regulated by a group of genes (the *tasselseed* loci). In dioecious plants there are several mechanisms of sex determination. One is XY sex chromosomes and the ratio of these to the autosomes (in *Silene* and *Rumex*, for example).

Fertilization and formation of the zygote

The pollen tube grows through the micropyle into the ovule and contacts the embryo sac. The pollen tube's contents are emptied into one of the two synergids, which is then destroyed along with the vegetative pollen tube. The two sperm cells penetrate further into the embryo sac; one of them forces its way into the egg cell, and their combined nuclei fuse to form the diploid zygote nucleus. The other sperm cell migrates to the secondary embryo sac's diploid nucleus and fuses with it, forming a triploid endosperm nucleus. By this double fertilization, the diploid zygote is formed as the initial stage of the new sporophyte that will grow to form the embryo (section 6.1). At the same time the endosperm nucleus forms, together with the cytoplasm of the embryo sac, which is the basis for the triploid endosperm that will later surround the embryo. The integuments of the ovule form the testa that protects the contents of the maturing seed (that is, embryo plus endosperm). In parallel to seed maturation, the carpels differentiate to form the fruit.

While molecular genetic analyses of pollination have already yielded rich results (see above), there have been few reports on fertilization. However, it has recently become possible (for example in maize) to isolate sperm as well as egg cells and then fuse them (*in vitro* fertilization). From these, an intact plant has been regenerated. As well as its potential for breeding, this method can be anticipated to provide a key to a better understanding of development in this phase of the plant's life cycle.

6.4 Bibliography

Embryogenesis and postembryonic vegetative development

Aeschbacher, R. A., Schiefelbein, J. W., and Benfey, P. N. (1994). The genetic and molecular basis of root development. *Annual Review of Plant Physiology and Plant Molecular Biology*, **45**, 25–45.
Review of investigations on *Arabidopsis*.

Barton, M. K. and Poethig, R. S. (1993). Formation of the shoot apical meristem in *Arabidopsis thaliana*: an analysis of development in the wild type and in the *shoot meristemless* mutant. *Development*, **119**, 823–31.
Thorough description.

Clark, S. E., Running, M. P., and Meyerowitz, E. M. (1993). *CLAVATA1*, a regulator of meristem and flower development in *Arabidopsis*. *Development*, **119**, 397–418.
Shows that changes to the meristem also affect flower gene expression.

Dolan, L. *et al.* (1993). Cellular organisation of the *Arabidopsis thaliana* root. *Development*, **119**, 71–84.
Basis for further work.

Hülskamp, M., Miséra, S., and Jürgens, G. (1994). Genetic dissection of trichome cell development in *Arabidopsis*. *Cell*, **76**, 555–66.
First detailed description of trichome development.

Irish, V. F. and Sussex, I. M. (1992). A fate map of the *Arabidopsis* embryonic shoot apical meristem. *Development*, **115**, 745–53.
Uses clonal analysis.

Jackson, D., Veit, B., and Hake, S. (1994). Expression of maize *KNOTTED1* related homeo box genes in the shoot apical meristem predicts patterns of morphogenesis in the vegetative shoot. *Development*, **120**, 405–13.
Use of *in situ* hybridization and immunostaining.

Johri, B. M., Ambegaokar, K. B., and Srivastava, P. S. (1992). *Comparative embryology of angiosperms*. Springer, Berlin.
Very detailed comparison.

Jürgens, G. and Mayer, U. (1994). *Arabidopsis*. In *Embryos. Color atlas of development*, (ed. J. B. L. Bard), pp. 7–21. Wolfe, London.
A detailed partitioning of embryogenesis into 20 phases using morphological criteria as bases for analysis of mutant phenotypes.

Jürgens, G., Torres Ruiz, R. A., and Berleth, T. (1994). Embryonic pattern formation in flowering plants. *Annual Review of Genetics*, **28**, 351–71.
Review article.

Malamy, J. E. and Benfey, P. N. (1997). Organization and cell differentiation in lateral roots of *Arabidopsis thaliana*. *Development*, **124**, 33–44.

The origin of tissue layers in the developing lateral root has been studied in transgenic plants that express β-glucuronidase (GUS) under the control of tissue-specific regulatory sequences.

Mayer, U., Torres Ruiz, R. A., Berleth, T., Miséra, S., and Jürgens, G. (1991). Mutations affecting body organization in the *Arabidopsis* embryo. *Nature*, **353**, 402–7.
First characterization and identification of embryogenic pattern mutants in plants.

Meyerowitz, E. M. (1997). Genetic control of cell division patterns in developing plants. *Cell*, **88**, 299–308.
In this review, plant development is discussed as a problem of regulating cell division in time and space.

Scheres, B. *et al.* (1994). Embryonic origin of the *Arabidopsis* primary root and root meristem initials. *Development*, **120**, 2475–87.
The first clonal analysis.

Schwartz, B. W., Yeung, E. C., and Meinke, D. W. (1994). Disruption of morphogenesis and transformation of the suspensor in abnormal *suspensor* mutants of *Arabidopsis*. *Development*, **120**, 3235–45.
Describes mutants whose suspensors can also develop to form embryos.

Sinha, N., Williams, R., and Hake, S. (1993). Overexpression of the maize homeobox gene, *KNOTTED-1*, causes a switch from determinate to indeterminate cell fates. *Genes and Development*, **7**, 787–95.
An ectopically expressed gene is shown to be sufficient for the formation of the shoot meristem.

Steeves, T. A. and Sussex, I. M. (1989). *Patterns in plant development*, (2nd edn). Cambridge University Press, Cambridge.
New addition to a classical theory.

Torrcz Ruiz, R. A. and Jürgens, G. (1994). Mutations in the *FASS* gene uncouple pattern formation and morphogenesis in *Arabidopsis* development. *Development*, **120**, 2967–78.
First time that pattern formation and morphogenesis are described as distinct processes in plants.

Tsuge, T., Tsukaya, H., and Uchimiya, H. (1996). Two independent and polarized processes of cell elongation regulate leaf blade expansion in *Arabidopsis thaliana* (L.) Heynh. *Development*, **122**, 1589–600.
Mutant analysis suggests that the shape of the mature leaf results from genetically controlled directional cell elongation.

Van den Berg, C., Willemsen, V., Hage, W., Weisbeek, P., and Scheres, B. (1995). Determination of cell fate in the *Arabidopsis* root meristem by directional signalling. *Nature*, **378**, 62–5.
Cell-ablation experiments provide evidence that the fate of newly formed root cells is not determined by their clonal origin from the meristem initials but imposed by signals from the mature root tissues.

The following reviews were published in a special issue on 'Plant vegetative development' of *The Plant Cell* (July 1997) and give comprehensive overviews of current knowledge:

Clark, S. E. (1997). Organ formation at the vegetative shoot meristem. *The Plant Cell*, **9**, 1067–76.

Kerstetter, R. A. and Hake, S. (1997). Shoot meristem formation in vegetative development. *The Plant Cell*, **9**, 1001–10.

Larkin, J. C., Marks, M. D., Nadeau, J., and Sack, F. (1997). Epidermal cell fate and patterning in the leaves. *The Plant Cell*, **9**, 1109–20.

Laux, T. and Jürgens, G. (1997). Embryogenesis—a new start in life. *The Plant Cell*, **9**, 989–1000.

Poethig, R. S. (1997). Leaf morphogenesis in flowering plants. *The Plant Cell*, **9**, 1077–87.

Schiefelbein, J. W., Masucci, J. D., and Wang, H. (1997). Building a root: the control of patterning and morphogenesis during root development. *The Plant Cell*, **9**, 1089–98.

Flower formation

An, G. (1994). Regulatory genes controlling flowering time or floral organ development. *Plant Molecular Biology*, **25**, 335–7.
Short review on genetic aspects of flower development.

Becraft, P. W. (1995). Intercellular induction of homeotic gene expression in flower development. *Trends in Genetics*, **11**, 253–5.
Emphasis on cell–cell interactions during flower development.

Bowman, J. (1993). Arabidopsis – *an atlas of morphology and development*. Springer, Berlin.
Extremely comprehensive microscopic and photographic details of this important model plant.

Coen, E. S. (1996). Floral symmetry. *EMBO Journal*, **15,** 6777–88.
A brilliant account of pioneering work in the field.

Coen, E. S. and Meyerowitz, E. M. (1991). The war of whorls: genetic interactions controlling flower development. *Nature*, **353**, 31–7.
A classical description of the ABC model of flower development.

Coupland, G. (1995). Genetic and environmental control of flowering time in *Arabidopsis*. *Trends in Genetics*, **11**, 393–7.
Summarizes new work in this area.

Drews, G. N. and Goldberg, R. B. (1989). Genetic control of flower development. *Trends in Genetics*, **5**, 256–61.
One of the first reviews to focus on mutants in flower development.

Gasser, C. S. (1991). Molecular studies on the differentiation of floral organs. *Annual Review of Plant Physiology and Plant Molecular Biology*, **42**, 621–49.
Molecular classification of flower-specific genes and proteins.

Goldberg, R. B. and Chasan, R. (1993). *Plant reproduction*. American Society of Plant Physiologists, Rockville, MD.

Huala, E. and Sussex, I. M. (1993). Determination and cell interactions in reproductive meristems. *The Plant Cell*, **5**, 1157–65.
Overview of reproductive meristem development.

Huijser, P. and Klein, J. (1994). Neue Erkenntnisse zur Blütenentwicklung. *Biol. uns Zt*, **24**, 21–9.
Review summarizing flower development (German text). Nice figures.

Jürgens, G. (1997). Memorizing the floral ABC. *Nature*, **386**, 17.
News and views article on key results in flower development.

Ma, H. (1994). The unfolding drama of flower development: recent results from genetic and molecular analyses. *Genes and Development*, **8**, 745–56.
Extends the work of Coen and Meyerowitz.

Ma, H. (1997). The on and off of floral regulatory genes. *Cell*, **89**, 821–4.

Schwarz-Sommer, Z. *et al.* (1992). Characterization of the *Antirrhinum* floral homeotic MADS-box gene *deficiens*: Evidence for DNA binding

and autoregulation of its persistent expression throughout flower development. *EMBO Journal*, **11**, 251–63.
Functional proof of a regulatory gene: illustrates the techniques used.

Van der Krol, A. R. and Chua, N.-H. (1993). Flower development in petunia. *The Plant Cell*, **5**, 1195–203.
Demonstrates the virtues of *Petunia* as a model system for the genetic analysis of flowering.

Weberling, F. (1989). Morphology of flowers and inflorescences. Cambridge University Press, Cambridge.
A systematic presentation of flower morphology.

Weigel, D. (1995). The genetics of flower development: From floral induction to ovule morphogenesis. *Annual Review of Genetics*, **29**, 19–39.
Comprehensive and up-to-date review of many aspects of flower development.

Weigel, D. and Meyerowitz, E. M. (1993). Activation of floral homeotic genes in *Arabidopsis*. *Science*, **261**, 1723–6.
Very clear results from *in situ* hybridization.

Weigel, D. and Meyerowitz, E. M. (1994). The ABCs of floral homeotic genes. *Cell*, **78**, 203–9.
One of the core reviews in the field. Short review of all aspects of gene regulation in the flower. Includes discussion of analogies to animal development systems.

Yanofsky, M. F. (1995). Floral meristems to floral organs: genes controlling early events in *Arabidopsis* flower development. *Annual Review of Plant Physiology and Plant Molecular Biology*, **46**, 167–88.
In-depth review of early aspects of flowering

Gametophyte development

Dodds, P. N., Clarke, A. E., and Newbigin, E. (1996). A molecular perspective on pollination in flowering plants. *Cell*, **85**, 141–4.
Focus on molecular-genetic aspects of pollination.

Drews, D. H. and Goldberg, R. B. (1989). Genetic control of flower development. *Trends in Genetics*, **5**, 256–61.
Clear introductory article.

Gasser, C. S. and Robinson-Beers, K. (1993). Pistil development. *The Plant Cell*, **5**, 1231–9.
Review of work on pistil development.

Goldberg, R. B., Beals, T. P., and Sanders, P. M. (1993). Anther development: basic principles and practical applications. *Plant Cell*, **5**, 1217–29.
Short review of the basics of anther development.

Hülskamp, M., Schneitz, K., and Pruitt, R. E. (1995). Genetic evidence for a long-range activity that directs pollen tube guidance in *Arabidopsis*. *The Plant Cell*, **7**, 57–64.
A detailed study of pollen tube development.

Klucher, K. M., Chow, H., Reiser, L., and Fischer, R. L. (1996). The AINTEGUMENTA gene of arabidopsis required for ovule and female gametophyte development is related to the floral homeotic gene APETALA2. *The Plant Cell*, **8**, 137–53.
Focus on the female gametophyte.

McCormick, S. (1993). Male gametophyte development. *The Plant Cell*, **5**, 1265–75.
Good extension to Goldberg's article.

Mascarenhas, J. P. (1989). The male gametophyte of flowering plants. *The Plant Cell*, **1**, 657–64.
Focus on the male gametophyte.

Mascarenhas, J. P. (1993). Molecular mechanisms of pollen tube growth and differentiation. *The Plant Cell*, **5**, 1303–14.
Review of pollen tube development.

Nasrallah, J. B. and Nasrallah, M. E. (1993). Pollen–stigma signaling in the sporophytic self-incompatibility response. *The Plant Cell*, **5**, 1325–33.
Summarizes work in this area.

Newbigin, E., Anderson, M. A., and Clarke, A. E. (1993). Gametophytic self-incompatibility systems. *The Plant Cell*, **5**, 1315–24.
Summarizes work in this area.

Reiser, L. and Fischer, R. L. (1993). The ovule and the embryo sac. *The Plant Cell*, **5**, 1291–301.
Review article.

Reiser, L. *et al.* (1995). The BELL1 gene encodes a homeodomain protein involved in pattern formation in the *Arabidopsis* ovule primordium. *Cell*, **83**, 735–42.
One of the few well-characterized genes involved in ovule development.

7 Pathogens and symbionts as growth modulators

Contents

7.1 **Viruses and viroids**

7.1.1 Symptoms caused by plant viruses differ widely

7.1.2 Viruses and viroids have simple structures

7.1.3 Plant viruses have very compact genomes

7.1.4 Plant viruses move from cell to cell via the plasmodesmata

7.1.5 How plant viruses influence plant development

7.1.6 Viruses can protect plants from other viruses

7.1.7 Viruses cause specific plant reactions that can lead to resistance

7.2 **Agrobacteria**

7.2.1 The phenomenon

7.2.2 Genetic analysis of tumour generation

7.2.3 Agrobacteria as gene vectors

7.3 **Rhizobia**

7.3.1 What can be seen?

7.3.2 Genetic analysis

7.3.3 Cross-talk between symbiont and host

7.3.4 Establishment of symbiosis

7.4 **Bibliography**

Preface

In nature, plant development is regulated by more than just endogenous and abiotic factors. In many cases, interactions between different organisms are very important. Fungi, bacteria, and viruses can have a major impact on plant morphogenesis, and their interactions with plants can be either beneficial or harmful—judged by the overall fitness of a plant population within a given ecosystem, rather than by how individual plants fare.

If we are to examine the molecular basis of these interactions between different organisms, we have to disregard the organisms' functions in their ecosystem. It therefore makes sense to regard pathogens and symbionts as 'modulators of development'. In recent years, the molecular prerequisites for the coexistence of plants and micro-organisms have been studied intensively. Special importance has been placed on a number of model systems, which have involved plant viruses, *Agrobacterium tumefaciens*, and rhizobia. These examples have given us special insights into the mechanisms of coexistence.

Plant viruses are suitable organisms to use for the investigation of these interactions because of their small genome sizes (making it relatively easy to retain an overview of their genetics). *Agrobacterium* has become famous as nature's own genetic engineering system for plants, and its abilities can be exploited readily. Rhizobia attracted attention because of their ability to supply certain plant species with nitrogen from the atmosphere.

In recent years, each of these three systems has been shown to have its own unique features. However, as knowledge has increased, it has become clear that there are general mechanisms between the host and micro-organism that remain valid for many such interactions. When a foreign organism enters a plant, the plant reacts at first with very general, non-specific defensive reactions, with the *de novo* synthesis of PR-proteins (pathogenesis-related proteins). Then secondary plant metabolites, such as flavones and their derivatives, are produced. For *Agrobacterium* and rhizobia, these flavonoids act as signals to induce bacterial genes specifically involved in the bacterium–plant interaction.

The second step is recognition. Both organisms must recognize each other, and at this step it becomes apparent whether the plant is resistant or susceptible to the invading micro-organism. An important recent discovery was that plants use a similar mechanism to recognize both viruses and bacteria. The third defensive step, which many different organisms have in common, is the hypersensitive reaction. Infected plant tissues 'commit suicide' in order that the rest of the plant can survive, because with the death of its host tissue, the invading micro-organism is no longer able to spread. In all probability, the next few years will show that regardless of the type of micro-organism that induced them, the same host signal–transduction chains are responsible for inducing the hypersensitive response.

Other than the effects mentioned above, some of the most remarkable interactions are with the host plant's hormone system. For instance, agrobacteria can interfere genetically with the balance of auxin and cytokinin, while rhizobia produce substances (called *nod*-factors) that imitate plant hormones, so that they can manipulate the plant's growth to their own ends. Because of this interplay, it is hoped that studies into plant–micro-organism interactions will yield not only information about symbiosis and pathogenesis, but also some valuable insights into the developmental processes of plants.

7.1 Viruses and viroids

Plant viruses are usually simple combinations of a nucleic acid core and a protein coat. Very few plant virus species are enveloped within a membrane. Viroids have neither a protein coat nor a membrane, and simply consist of a circular RNA molecule. Viruses and viroids are (genetically) autonomous units but, because they lack a metabolic apparatus of their own, they are entirely dependent on the host plant for replication, hence their characterization as obligate parasites. However, this characterization may be a little unfair, and has arisen because plant virologists are largely concerned with viruses that cause diseases. As a result, the majority of described plant viruses are pathogenic towards plants. However, the truth is that plant viruses are much more widespread than the occurrence of macroscopic symptoms might lead us to believe. Only since the development of molecular methods has it been shown that both crop and wild plants contain an extraordinary variety of viruses. Such viruses are destined to remain unrecognized if they do not have any visible or measurable effect on the development of their host plant.

Judging whether a certain trait caused by viral infection is symptomatic of disease depends very much on your point of view. In agriculture, plant stunting can be a disadvantage, but to a horticulturist it can be highly desirable.

And how can these symptoms be evaluated in different ecosystems? In windy conditions at the seaside or on a mountainside, stunting might increase the fitness of a population of plants by preventing them from blowing over.

Ignoring the scientific interest and judgement of virologists for the moment, we are left considering viruses as genetically semi-autonomous units that, in one way or another, modulate plant development. In fact, historically the first plant virus infection recognized as such was one that caused a highly desirable trait in a horticultural plant. In AD 752, a Japanese poem from the Manyoshu anthology described the plant species *Eupatorium chinense*, and the description now makes it plain that it was in fact infected with a geminivirus; seventeenth-century Dutch painters enthusiastically painted flaming tulips (infected with a potyvirus); and, finally, Abutilon mosaic virus was spread all over the world during the nineteenth century because the marbling pattern produced in infected leaves made these plants more valuable (Figure **7.1**).

Pathogenicity is therefore not an intrinsic feature of plant viruses, but is merely the result of the type of interaction of a specific host genotype with a specific virus strain. Moreover, the occurrence of disease is dependent on a whole network of environmental factors. Plant viruses that are horizontally transmitted may also bestow on their host plants an additional epigenic adaptational ability, just as bacteria can benefit from being infected with certain plasmids and bacteriophages.

7.1.1 Symptoms caused by plant viruses differ widely

Definition The word *symptom* comes from the Greek συμπτωμα (chance, transient peculiarity). The meaning has changed to 'a feature caused by a disease'.

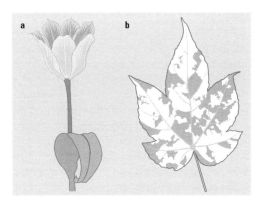

Figure 7.1 Classical examples of virus-infected plants. (a) 'Flaming tulips', the flower patterns are the result of the actions of a potyvirus; (b) marble-like mosaic on an *Abutilon* leaf that is infected with the Abutilon mosaic geminivirus.

Viral symptoms are classified as: changes in colour, changes in shape, and necrosis (death of the whole plant or parts of it).

Changes in colour can affect different organs differently. In most cases, leaves become yellow, red or dark green, or the colours of the flowers change. The leaves can be striped, have a mosaic pattern, or localized chlorotic spots. Ringspots are special cases that consist of concentric rings. The veins can be darker or lighter than the interveinal regions.

A change in shape includes symptoms such as leaf rolling, stunting, leaves becoming fern-like, tumours, or teratomes. The plant's habit can be altered: internode length can be shortened, growth can become bushy, roots can proliferate (rhizomania), or the number and size of flowers and fruits can be altered. Necrosis can either be local and lead to local lesions or can affect the entire plant.

Symptoms depend more on the time of infection rather than being fixed features of particular viruses. Depending on the susceptibility of a plant to a given virus, the leaf veins may lighten; later during the plant's development a mosaic pattern could form, and finally the plant may be killed. Symptoms can also be temporary, in that there is always the chance that the plant could recover from the infection. In general, viruses infect young organs in young plants most readily. Many viruses utilize the dynamic growth spurts of younger plants in order to replicate.

Symptoms can also depend largely on environmental conditions (for example, light, temperature, and day length) and the plant's own internal conditions (for example, hormone concentration, mitotic activity, or morphological differentiation).

Definition By convention, plant virus species are named (in order) after (1) the plant they were first isolated from (for example, tobacco), (2) the symptoms on this plant (for example, mosaic); and (3) the virus genus (for example, tobamovirus) if it has been assigned to one, or just 'virus' otherwise.

This naming scheme can be somewhat arbitrary and very often confusing. For example, if tobacco mosaic virus (TMV) had first been isolated from *Nicotiana tabacum* cv. Samsun *NN*, it would have been called 'tobacco local necrosis virus', because on this particular plant, TMV causes localized necrotic spots instead of mosaic symptoms. As with virus pathogenicity, this example shows that the type of symptoms produced is not an inherent virus feature, but is instead the result of a virus–host interaction. Symptoms depend on the genotypes of both partners. Whether symptoms are expressed only locally at the site of infection or systemically throughout the whole plant can depend on the genetic disposition of the plant and on the plant's environment.

Besides an accumulation of virus particles, changes at the cellular level are characterized mainly by changes to the plastids. Changes such as starch accumulation, degradation, and restructuring of the thylakoids, and an increase in the number of plastoglobuli can occur. The cell wall may thicken and the size of the plasmodesmata may change. Anthocyanins may accumulate in the vacuole, changing the leaf colour to red. Surprisingly, the nucleus and the mitochondria are seldom changed (if we ignore virus particle accumulation in the nucleus). Often, the cytoplasm will contain crystallized or amorphous inclusions (inclusion bodies) of virus-encoded proteins.

7.1.2 Viruses and viroids have simple structures

In contrast to the complexity of symptom expression, the structure of most viruses is relatively simple. The overwhelming majority of known plant viruses consist of one or more nucleic acid molecules (usually RNA, seldom DNA) and a coat protein. Viroids do not even possess a coat protein and their RNAs are only a few hundred bases long, but nevertheless they can still cause a wide variety of symptoms, from minor shape changes to the death of coconut palm trees.

The simplicity of the viral coat is made possible by two structures—the helix and the icosahedron. Both can yield space-filling constructions using only a single type of protein building block. The range of shapes that can result from these basic structures is shown in Figure **7.2**.

7.1.3 Plant viruses have very compact genomes

Plant viruses are not just small, they also have very compact genomes (Figure **7.3a–e**). An archetypal minimal virus genome would contain genes to encode a replication protein (RP), a movement protein (MP) to move the virus throughout the plant, and a coat protein (CP). If the virus is transmitted by a vector (usually an insect), it may have another gene dedicated to vector transmission (VP). Usually, viral genes are arranged linearly in the order (starting at the 5' end): replication, movement, coat (for example, TMV; Figure **7.3a**). With this arrangement, each gene's function as early or late is fixed. Of course, there are many variations on this basic theme, most of which have developed as adaptations to multiplication requirements.

Figure 7.2 Samples of plant virus structures. Most plant viruses contain single-stranded (ss) RNA in (+) sense orientation. Only a few, such as reoviruses and cryptoviruses, package double-stranded (ds) RNA. Three groups (caulimo-, badna-, and geminiviruses) contain DNA. The virus particle structure has two basic forms: the icosahedron and the helix. A helical structure can form a rigid rod (tobamoviruses) or a flexible rod (potyviruses). Some groups also form bacilliform structures (badnaviruses, alfalfa mosaic virus, the nucleocapsid of rhabdoviruses). The geminiviruses are unique in that they are composed of two incomplete icosahedrons. With the exception of the rhabdo- and tospoviruses, plant viruses lack a membranous envelope. (According to Matthews 1991.)

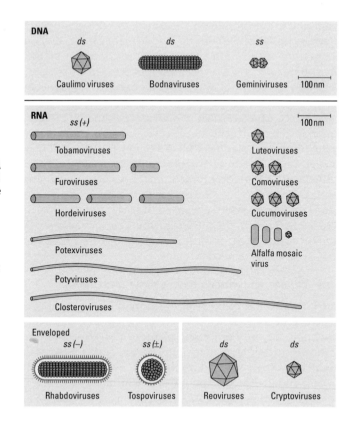

Of all the plant viruses described in the literature, 94 per cent have RNA genomes, and only 6 per cent have DNA. The majority (82 per cent) of RNA viruses have linear positive-sense single-stranded RNAs.

Definition The positive/negative designation for RNA virus genomes is defined as follows: if the virion RNA can be translated directly by ribosomes (as if it were mRNA), it is said to be positive sense. If a complementary strand must be made before translation can take place, the RNA is *negative sense*.

Viral RNAs face a problem that was described by Marilyn Kozak for mRNAs in eukaryotic cells (the Kozak rule). According to this rule, eukaryotic ribosomes move along an mRNA, starting at the 5′ end, until they come across the first usable start codon (AUG in certain sequence contexts). Here the ribosome starts the translation process, and it stops at the first stop codon it comes to in the frame. Any start codons following this stop codon are ignored. Plant viruses have developed several different strategies to work around this limitation (Figure **7.4**):

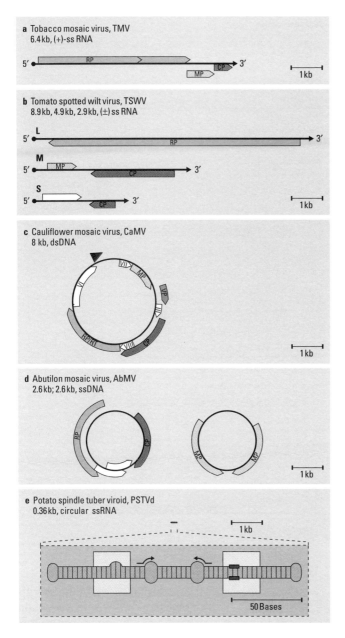

a Tobacco mosaic virus, TMV
6.4 kb, (+)-ss RNA

b Tomato spotted wilt virus, TSWV
8.9 kb, 4.9 kb, 2.9 kb, (±) ss RNA

c Cauliflower mosaic virus, CaMV
8 kb, dsDNA

d Abutilon mosaic virus, AbMV
2.6 kb; 2.6 kb, ssDNA

e Potato spindle tuber viroid, PSTVd
0.36 kb, circular ssRNA

Figure 7.3 Genome structures of selected plant viruses. Only a few genes are necessary for the life cycle of a plant virus: RP, replication proteins; MP, movement protein (the protein for transport within the plant); CP, capsid protein (for transmission between plants); VP, vector protein (for specific transmission by vectors, for example fungi or insects). The genome can consist of RNA (TMV, TSWV) or of DNA (CaMV, AbMV). The genome of CaMV is special because it transcribes its DNA genome into genomic RNA (35S RNA) which is re-transcribed into DNA in a reverse transcription reaction. Therefore CaMV is called a *pararetrovirus* in reference to the animal retroviruses. The genome of plant viruses can be monopartite (TMV, CaMV), bipartite, tripartite, and so on (multipartite). Viroids (PSTVd) differ from plant viruses in that their circular RNA genomes apparently have no protein-coding capacity. Their single RNA strand can form a very stable rod-like structure by base-pairing. (According to Matthews 1991.)

1. Creation of subgenomic RNA(s). The viral RNA can have promoter structures which the viral RNA-dependent RNA polymerase can transcribe as additional subgenomic RNAs, each having a start codon close to its 5′ end.

2. Multipartite genomes. The viral genome is divided into two or more RNAs that are packaged in one or more

Figure 7.4 Strategies for the expression of compact viral genomes (for an explanation see text). (According to Matthews 1991.)

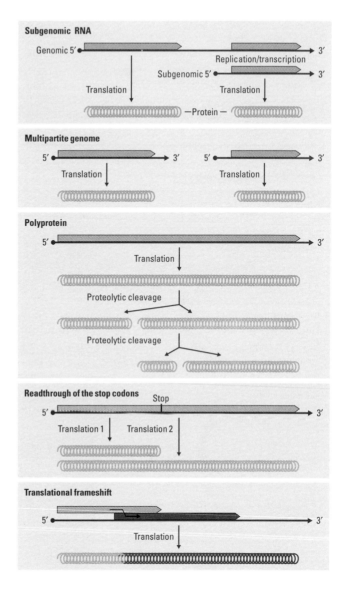

virus particles. In most cases, all particles need to be present in a cell to be infectious. This arrangement differs from (1) because each genome part differs from the others and is a unique template sequence for its own replication. Subgenomic RNAs are not packaged (although they might be by chance).

3. Polyproteins. Some viral RNAs are first transcribed into a single, long polyprotein, which is later processed into many functional proteins by virus-encoded proteinases.

4. Read-through. Some proteins are translated by suppression of the first stop codon. This results in a fusion protein that can have a different function at its C-terminal end.

5. Translational frameshift. In a few cases it has been shown that ribosomes change to a different reading frame during translation, resulting in a new protein.

Besides these strategies to work around Kozak's rule, some enterprising DNA viruses have managed to break it altogether and create polycistronic mRNA (cauliflower mosaic virus; Figure **7.3c**).

7.1.4 Plant viruses move from cell to cell via the plasmodesmata

In contrast to animal viruses, plant viruses move *after* they have entered the host organism's symplasm. The cytoplasm of each cell is connected to its neighbours by plasmodesmata (Figure **7.5**). However, viruses cannot move freely through plasmodesmata, which must first be widened. Plant viruses possess special proteins to do this, called movement proteins (MP) (Box **7.1**; cf. Figure **7.3**). Viruses can spread throughout a plant in at least two ways: (1) as nucleic acid–movement protein complexes (e.g. TMV); or (2) as a virion within a tubule made by the movement protein.

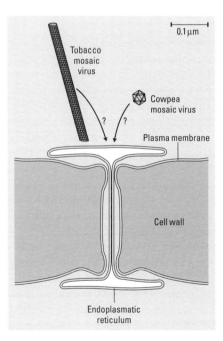

Figure 7.5 Comparison of the sizes of selected plant viruses (tobacco mosaic virus, cowpea mosaic virus) to the sizes of plasmodesmata. In order for the plant viruses to fit through the plasmodesmata, the plasmodesmata must widen and the viral nucleic acid must stretch. (According to Gibbs 1976.)

Box 7.1 Cell-to-cell movement of tobacco mosaic virus

A well-known example of a movement protein (MP) is the 30 kDa protein of TMV (*see* Figure **7.3**). If the gene for this is mutated, the virus cannot move between cells. Such a defect can be overcome by the intact movement proteins from other co-inoculated 'healthy' viruses, or from transgenic movement protein expressed by the plant. Immunological methods can prove that the movement protein is located in and around the plasmodesmata. The function of *Escherichia coli*-expressed movement protein can be tested. Fluorescently labelled dextran beads can determine the exclusion size of the plasmodesmata. If the movement protein is co-injected with these dextran beads, the exclusion size of the plasmodesmata is raised from 1.5 nm to between 6 and 9 nm.

However, if the viral RNA was folded and tangled with an extensive secondary structure, it would be unable to squeeze through even the enlarged pores. To get through the pore, it must be stretched into a long, straight filament, then 'threaded' through the pore. As well as enlarging the plasmodesmata, the movement protein can 'stretch' the RNA until its diameter is only 2 nm.

Movement proteins have a very low specificity towards the viral RNA, so many different virus species are able to complement each other's movement function. For example, a potato virus X with a mutated and useless movement protein can move from cell to cell if it is co-inoculated with wild-type TMV.

Other than the two methods mentioned above, there is another way to ensure cell-to-cell movement. An example of the morphogenetic potential of some viruses is given by a family of plant viruses, whose best-known member is cowpea mosaic virus. When the movement protein, or whole virions, of viruses belonging to this family is injected into protoplasts, the plasma membrane is induced to build external tubular extensions. These tubules have also been seen in whole plant cells, where they span the plasmodesmata. In both plant and protoplasts, these tubules contain whole virions.

7.1.5 How plant viruses influence plant development

With the exception of plasmodesmata modification, there are currently no other known instances of viruses specifically interfering with specific plant structures or processes. For human and animal viruses, especially in cases of tumour development, many specific interactions between viral and cellular proteins have been observed, which, via a signal transduction chain, interfere with control of the cell cycle. Such protein–protein interactions have yet to be shown in plant viruses. Initial evidence for these interactions will be discussed in section 7.1.7.

At least four explanations could account for the indirect pathogenic effects of viruses:

1. A simple explanation for virus symptoms might be that virus replication occurs at the expense of the host. However, there is a paradoxical situation whereby in some virus–plant interactions enormous virus titres are observed, with no apparent adverse effect on the plant. Such plants, in which viruses can replicate without producing symptoms, are said to be tolerant.

2. Another simple assumption might be that virus-induced changes to the plasmodesmata can harm symplastic transport. However, there is no proof for this assumption, as transgenic plants containing viral movement proteins and widened plasmodesmata seem to develop perfectly normally.

3. A third assumption might be that virus proteins interact by regulating host-gene expression. Many viral proteins bind to non-viral RNA or DNA. However, no specific interactions of host nucleic acid and viral proteins have been shown to date.

4. The effects of viruses upon plant hormone metabolism have been investigated, because many symptoms caused by plant viruses are similar to those caused by plant hormones. However, there is a wide range of effects here. Viruses can raise or lower the concentration of a particular plant hormone, or change the balance of different plant hormones. For these reasons it becomes rather questionable whether viruses influence plant hormone concentrations directly, or whether the observed effects are merely the result of altered plant development (Box **7.2**).

In summary, one of the most fascinating challenges in the future will be to investigate the interactions between plant viruses and their hosts that lead to the development of symptoms.

A special case for symptom development is that resulting from infection with viroids or satellite viruses. As far as viruses are concerned, it seems a reasonable assumption that virus proteins are responsible for their effects on the plant. In the case of viroids, this is impossible, and for satellite viruses the effects of proteins have been excluded

---

Box 7.2 Experiments to help determine how symptoms are generated

Two methods can be employed to discover the cause of symptoms. One way is to mutate individual virus genes with site-directed mutagenesis, and then observe the effect (if any) on symptom development. The second method involves making transgenic plants that express viral proteins, then studying the effects. The first type of experiments has tended to show many different effects, with no apparent links between them. One obvious problem is that by altering the viral genes, one may alter its ability to replicate or move, as well as altering symptom production.

The second approach is usually used to achieve an entirely different aim—to protect plants from virus infection. This is because in many cases it has been shown that if the plant has single viral genes from a particular virus species, introduced as transgenes, the plant will become resistant to infection by that virus species. It is usually the exception, rather than the rule, that these transgenes lead to the development of symptoms of viral disease.

experimentally. In both cases, scientists are looking for a direct role of the viroid/satellite virus RNA as an effector.

Definition Viroids are RNA molecules which are not associated with proteins, and which do not have any protein-coding capacity (Figure **7.3e**). They are circular RNAs that, via intramolecular base-pairing, become helical or rod-shaped.

This base-pairing is not only necessary for the stability of the RNA, but also affects symptom production. Mutagenesis experiments have shown that only a few nucleotides need be changed to turn a moderately virulent into a highly virulent viroid, and vice versa. The thermodynamically stable RNA structure of the viroid seems to be the cause of pathogenesis, and not any coding capacity of the RNA.

Definition Satellite viruses and satellite RNAs are defined as such by their inability to replicate without a helper virus. Satellite viruses differ from satellite RNAs in that they encode their own coat protein; satellite RNAs are encapsidated with the coat protein of their helper virus.

The nucleic acid sequences of both types of satellite differ from that of their helper viruses. Satellite viruses and RNAs were discovered because they modulate symptoms caused by their helper viruses (Box **7.3**).

In summary, it can be said that in the cases of viroids and satellite RNAs, symptoms are not only a consequence of a complex interaction between viruses, satellites, host plants, and the environment, but also that the RNA structures themselves are able to modulate morphogenesis.

7.1.6 Viruses can protect plants from other viruses

Definition If a plant is first inoculated with a mild strain of a virus, and then later on with a more virulent related strain, replication of the more virulent strain can be prevented (Figure **7.7**). This effect is known as cross-protection.

This cross-protection can be established by both plant viruses and satellite RNAs (in conjunction with their helper virus).

As yet, this phenomenon remains unexplained. With the advent of gene technology, it seemed that explaining the process would be straightforward. Single viral genes were isolated and integrated into plants' genomes, where they were expressed. These experiments showed that production of the viral coat protein alone by the plant was sufficient to achieve protection. These results were interpreted as meaning that invading viruses were unable to remove their coat proteins, and therefore could not replicate.

Box 7.3 Cucumber mosaic virus—an example of a helper virus with its satellite RNAs

Cucumber mosaic virus (CMV) and its satellite RNAs provide an excellent example of the interaction between helper viruses and satellite RNAs (Figure 7.6). CMV has the widest host range of all known plant viruses. Hosts for CMV include 775 plant species, of 365 genera from 85 families. Different CMV strains cause different symptoms on different host plants, and in the past this led to them being misclassified as different virus species. Only detailed molecular and immunological investigations have revealed that in reality there is only a single species. CMV has three genomic RNAs (Figure 7.6) and an additional subgenomic RNA. RNA1 and RNA2 are packaged in separate particles; RNAs 3 and 4 are packaged together into one particle. Under experimental conditions, three particles are required for infection, RNAs 1, 2, and 3. For some CMV isolates, different satellite RNAs have been found, because they were packaged in the virus particles. They are small, single-stranded RNAs, 332–386 nucleotides long. They cannot replicate autonomously— they use CMV as a helper to do this. More than 25 different satellite RNAs have been isolated and sequenced. They have between 70 and 99 per cent homology, and have similar secondary structures. Most satellite RNAs replicate well with their CMV helpers in Solanaceae, but poorly in Cucurbitaceae. Their presence can attenuate symptoms, or enhance them, depending on the host. Even though most satellite RNAs will weaken the symptoms caused by their helper virus in the majority of hosts, the very same satellite RNAs can cause drastic symptoms in certain other plant host species (e.g. tomato, pepper, and tobacco). Just a single nucleotide change can be enough to alter host-specific symptoms. If the symptoms are weaker, it is not necessarily because there is less CMV in the plant.

A satellite RNA will have a different effect with different helper viruses. CARNA 5 (CMV-associated RNA 5) causes lethal necrosis of tomato with two different CMV strains, but causes stunting when in combination with a third strain. Symptoms depend on three things: host plant, helper virus, and satellite RNA.

Using *in vitro* recombination of cloned satellite RNA, domains in the 3′ region of the RNA that are associated with the onset of necrosis can be identified. In this region, a simultaneous change of three nucleotides can make a necrogenic satellite RNA non-necrogenic. However, a temperature change of just 2.5 °C is enough to prevent this necrosis. Even though the satellite RNA has an open reading frame, there is no evidence that the polypeptide it encodes plays any part in necrosis of the host plant. In fact, it seems that the primary, secondary, or tertiary structure of the RNA itself could be responsible.

One particularly elegant method by which to study the effects of the various satellite RNAs independently of the helper virus is to use transgenic plants that produce satellite RNA. These plants do not develop symptoms. However, the expected symptoms appear if the plant is inoculated with the right helper virus.

Later it was demonstrated that other viral proteins could provide the plant with protection against viral infection, and, finally, it has been shown that some untranslatable parts of the virus genome protect against further infection. It seems that it does not matter whether this is expressed as sense or antisense by the plant. A big surprise was that non-viral genes could also lead to cross-protection. If

Figure 7.6 The genome structure of cucumber mosaic virus, CMV. The genome consists of three elements (RNAs 1–3). An additional subgenomic mRNA is transcribed from the complementary strand of RNA 3. CMV occurs with satellite RNAs that are not related to the RNA of the virus. (After Matthews 1991; Palukaitis *et al.* 1992.)

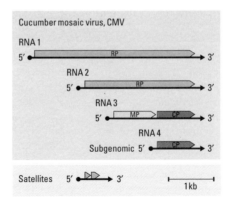

Figure 7.7 Principle of cross-protection. If plants are pre-inoculated with mild strains of TMV, they will be protected against more virulent virus strains. Similar protection can be achieved if just the coat protein gene of a TMV isolate is integrated transgenically into the plant's DNA.

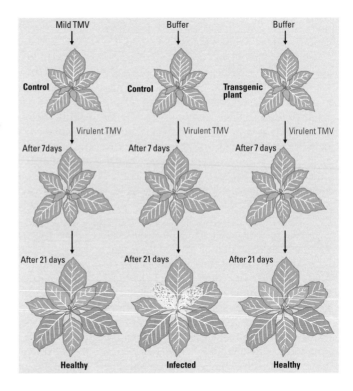

genetic engineering is used to introduce the gene for glucuronidase (GUS) into the potato virus X (PVX) genome, the chimeric virus can replicate and move normally within a host plant. Its infection can be monitored histochemically using a GUS-specific stain. But in a transgenic plant containing a GUS gene, the chimeric virus can neither replicate nor move (although the wild-type virus still can). Thus the homology between the virus gene and corresponding transgenes is important. There is yet another interesting effect: PVX only replicates within the plasma, leading to the assumption that interference between the two RNAs must be post-transcriptional, occurring after the viral RNA has been exported out of the nucleus.

These phenomena are similar to co-suppression, and perhaps viral systems can help us to understand this mechanism. Cross-protection may be explained by a number of different mechanisms, and so the term 'cross-protection' could, in fact, be a generic one, encompassing several such mechanisms.

7.1.7 Viruses cause specific plant reactions that can lead to resistance

Plants that are mechanically inoculated with a virus react by forming wounds. These wounds are similar to wound reactions that happen after infection with bacteria or fungi, or after damage due to other stress factors. The primary characteristic of this reaction is the production of pathogenesis-related proteins (PR proteins). These proteins can have a wide variety of functions. For example, they can be chitinases or glucanases, in spite of the fact that these do not have an immediately obvious function during a viral infection. Besides these general reactions to pathogens, there are some more specific ones. These effect only certain viruses or certain strains of viruses.

The classic example of such a specific reaction is the *N* gene-related resistance of some tobacco cultivars that can resist infection by certain TMV strains. *N* stands for necrosis-causing (for historic reasons, the term nc (necrotic) is used in a different cultivar). The *N* gene has been discovered in *Nicotiana glutinosa*, which cannot be infected systemically by TMV. Its spread is hindered by local necrotic lesions (Figure **7.8a,b**). Due to a hypersensitive reaction the invading

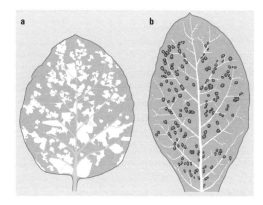

Figure 7.8 Different symptoms of TMV on two host species. (**a**) Systemic mosaic on *Nicotiana tabacum* Samsun *nn*; (**b**) local lesions on *Nicotiana sylvestris*. In (**b**) the infection is restricted to the inoculated leaf, while in (**a**) the entire plant is infected.

> ## Box 7.4 Finding the key to the *N* resistance genes
>
> Recently, the *N* gene was tagged by transposon mutagenesis, then isolated and sequenced. This gene was shown to be both necessary, and sufficient in itself, to start the *N*-specific hypersensitive reaction, by transferring it into *Nicotiana tabaccum* cv. Petite Havanna SRI, which normally does not show this type of reaction. Sequence comparisons lead us to believe that the *N* gene product is a protein that is part of an intracellular signal transduction chain. It has still to be shown exactly how and on what level the *N* gene product specifically recognizes TMV.
>
> The first signs seem to be that it is interacting somehow with the cell-to-cell movement of the virus. In trials with transgenic *NN* and *nn* plants that expressed the movement protein (MP) of TMV, only the *nn*
>
> plants had widened plasmodesmata (section 7.1.4) at room temperature. At higher temperatures, no hypersensitive reaction occurs. Both types of plants opened their plasmodesmata at 33 °C. Experiments on a different resistance gene have led to similar conclusions. The *tm1*, *tm2*, and *tm2²* genes confer resistance to tomato mosaic virus (ToMV), a close relative of TMV. Resistance-breaking strains of ToMV have been isolated. Molecular genetic analyses of their RNA sequences showed that the sequence changes responsible for breaking resistance are located in the MP gene. This interdependency of specific viral and host genes is a good 'real world' example of the gene-for-gene hypothesis that was developed for plant pathogens in a series of classical experiments (section 7.3.3).

viruses will be limited to the site of inoculation. The *N* gene is dominant and has been crossed into *Nicotiana tabacum*. The cultivar with the *N* gene is called *Nicotiana tabacum* Samsun *NN*; the one without is called *Nicotiana tabacum* Samsun *nn*. Although the spread of TMV is prevented in *NN* plants, other viruses can infect systemically without problems. Therefore this effect is virus specific (Box **7.4**).

7.2 Agrobacteria

Agrobacteria are considered to be unique because they evolved the ability to exchange genes between two different kingdoms, bacteria and plants. This genetic exchange is not merely by chance—the bacterium executes a deliberate and highly targeted natural molecular genetic alteration of infected plant cells. Agrobacteria make infected plants differentiate tissue to form tumours, in which they replicate. They change the plant's metabolism in such a way as to produce special amino acids which are used only by agrobacteria. Knowledge of their interactions with plants has many widespread implications, and has kick-started the plant genetic engineering industry. Agrobacteria have become useful tools that scientists use routinely to transfer a wide range of genes into plants.

7.2.1 The phenomenon

A susceptible plant which is injured in the presence of *Agrobacterium tumefaciens* (Rhizobiaceae) will produce tumours at the site of infection. Agrobacteria spread throughout a plant via the xylem. If the plant is wounded again at another site, secondary tumours can arise. In fact, *Agrobacterium tumefaciens* was named as such because of its tumour-inducing capability. *Agrobacterium radiobacter* cannot induce tumours. Unlike *Agrobacterium radiobacter*, *Agrobacterium tumefaciens* contains a large plasmid called the Ti plasmid (tumour-inducing plasmid), which has all the necessary genes for inducing tumours.

Agrobacteria require small wounds as entry points into a plant. At the same time, the wounding process makes the plant produce signals that alter the regulation of the bacterial genes. Once this has occurred, tumours will be induced within 18 hours. Infected plant tissue then differentiates and grows like a cancer. A special feature of these tumours is that in tissue culture they can grow independently of plant hormones. They also produce new kinds of amino acids called opines.

7.2.2 Genetic analysis of tumour generation

A bacterial plasmid of around 200 kb is needed to produce a tumour, and different *Agrobacterium* strains can carry different plasmids. They are classified according to the type of opines they make the plant synthesize—either nopaline or octopine (*see* Figure **7.11**). Nearly all of the Ti plasmid has been sequenced. Using techniques such as comparison with existing bacterial proteins, mutagenesis, and complementation tests, functions have been ascribed to most of the plasmid's genes. Two parts of all Ti plasmids have been found to be particularly important:

(1) the T-DNA: a part of the plasmid which is exported to the plant cell and is integrated into the cell's nucleus (Figure **7.9**); and

(2) the vir region, which encodes proteins involved in this transfer, but which stays within the bacterium (Figure **7.9**).

T-DNA is short for transferred DNA, and *vir* is an abbreviation of virulence. The term *vir* region might be considered misleading, as it has been demonstrated that the genes of this region are responsible for T-DNA transfer; thus it might be better to call it the *tra* region. However, *vir* region is the standard term in the literature, and it would only lead to further confusion to change it now. Both the vir

Figure 7.9 Principle of gene transfer from agrobacteria into plant cells. Ti, tumour-inducing plasmid; vir, *vir* region.

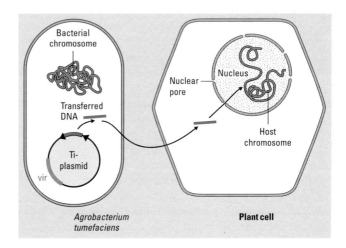

Figure 7.10 Genome structure of a Ti plasmid of the nopaline type of agrobacteria. T-DNA, transfer DNA; LB, left border; RB, right border; *vir*, *vir* region; *ori*, origin of replication; *tra*, conjugative transfer; *noc*, nopalin catabolism; *nos*, nopaline synthesis; *tms*, tumour morphology shoots; *tmr*, tumour morphology roots. (According to Koukolikova-Nicola *et al.* 1987.)

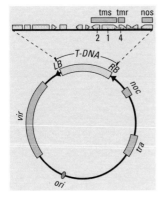

region and the T-DNA contain many genes with various different functions (Figures **7.10** and **7.12**).

Figure **7.10** shows the organization of the T-DNA, using as an example the nopaline plasmid. Only one sequence region of nopaline plasmids is transferred into the plant, whereas octopine plasmids transfer two. The region that is transferred is flanked by left (LB) and right (RB) borders, each consisting of 25 bp repeating sequences. These serve as markers for excision of the functional sequence between them.

How can T-DNA alter the way a plant cell differentiates? Mutation experiments have led to some interesting insights into the mechanisms involved. For example, if gene 4 is inactivated (Figure **7.10**), infected plants grow roots instead of tumours, hence gene 4 is named *tmr* (tumour morphology root). If genes 1 and 2 are inactivated, shoots develop instead of tumours (*tms*, tumour morphology shoot). Biochemical analyses of the products of these three genes have shown that they all influence the plant's hormone metabolism: *tms1* and *tms2* lead to the synthesis of the auxin, indole acetic acid, and *tmr* interferes with cytokinin synthesis. It has long been known that the ratio of auxins to cytokinins determines root and shoot morphogenesis (section 5.5.1). *Tmr* codes for an isopentenyltransferase enzyme that catalyses the synthesis of the cytokinin isopentenyladenosine-5′-monophosphate from dimethylallylpyrophosphate and 5′-AMP. *Tms1* and *tms2* are involved in a two-step synthesis of indole-3-acetic acid: the *tms1* gene product converts tryptophan into indole-3-acetamide which is subsequently processed by the *tms2* gene product to yield indole-3-acetic acid. Oncogenic func-

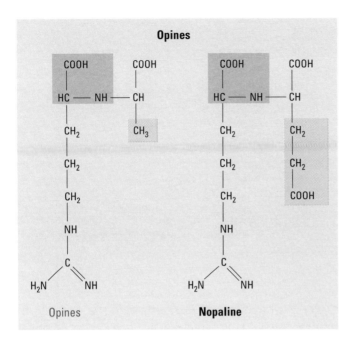

Figure 7.11 Chemical structure of opines.

tionality can thus be narrowed down to just three enzymes involved in hormone metabolism. Without these genes, agrobacteria would not be pathogenic; they would merely transfer their DNA without producing tumours. This is of vital importance when agrobacteria are used as gene vectors (section 7.2.3).

Further changes in the plant's metabolism result from the transfer of *nos* or *ocs* genes. Their enzyme products, nopaline synthetase and octopine synthetase, redirect amino acid metabolism (Figure **7.11**). The new enzymes synthesize nopaline and octopine, respectively. These are essential to the agrobacteria, but are of no use to the plant. T-DNA transfer can thus be viewed as natural genetic engineering of a target organism. In this case, the manipulating organism (the bacterium) not only creates for itself a suitable replication habitat (the tumour), it also makes the plant produce a ready source of nitrogen for its use (the opines).

The genes that interfere with the plant's hormone metabolism are necessary for tumour induction, but are not sufficient for pathogenity. For pathogenesis, many more chromosomal and plasmid genes are required. As well as the chromosomal genes (*chvA* and *chvB*) that are necessary for the bacteria to attach to plant cells, at the very least the genes from the *vir* region are essential for T-DNA transfer from bacterium to plant. Other plasmid genes are involved in determining the host specificity of the agrobacteria.

Figure 7.12 Functions of the *vir* genes. For explanation *see* Box 7.5 (according to Koukolikova-Nicola *et al.* 1987).

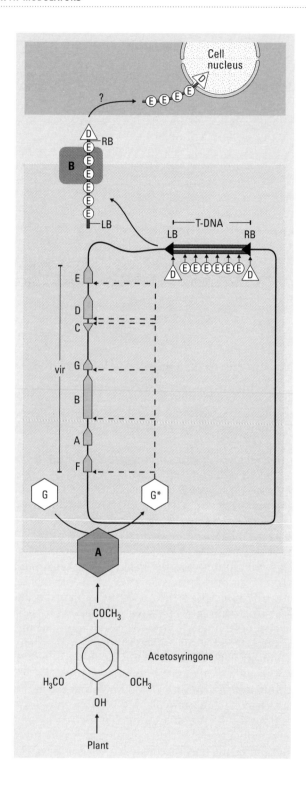

The functions of the *vir* region are well documented (Figure **7.12**)—its gene products participate in recognizing the wounded plant, direct the excision of the T-DNA and its subsequent movement through the bacterial plasma membrane, the cell wall, the plant cell's plasma membrane, and finally the pores of the cell's nucleus.

The vir region is 30 kb long and contains seven complementation groups (*virA* to *virH*) that can each be assembled by several genes; for example, *virD1* and *virD2* to make *virD*, or *virB1* through to *virB11* to assemble *virB* (Box **7.5**).

The T-DNA transfer process is reminiscent of bacterial conjugation, with the exception that the recipient organism is from the plant kingdom, where different promoters

---

Box 7.5 Transferring genes from agrobacteria to plants

Wounded plants excrete a variety of flavonoids, which agrobacteria recognize in order to activate their *vir* genes (Figure **7.12**). Acetosyringone has a similar effect, so it is used in experiments. Signal substances resulting from the wound reaction bind to the gene product of *virA*. VirA is anchored in the membrane and seems to activate VirG by phosphorylation. The VirG protein activates its own promoter and those of the other *vir* genes. In consequence, VirD2 and VirE2 proteins are produced. VirD2 is an endonuclease, which cuts specifically at the left and right borders (LB and RB). It links (probably covalently) to the 5' end of the bottom strand of RB in Figures **7.12** and **7.13**. VirE2 is a single-strand binding protein that sets free from the double strand the strand cut by VirD2. The resulting gap is filled by DNA repair mechanisms. The single-stranded T-DNA drifts as a complex with VirD2 and VirE2 to the bacterial membrane. It is not yet known how the complex crosses first the membrane, then the subsequent cell walls. However, there are signs that the 11 genes of *virB* encode a protein pore complex, which helps transfer the T-DNA/VirE2/VirD2 complex across the membrane. It is much clearer how the complex gets through the nuclear membrane. VirD2 and VirE2 have nuclear localization signals (NLS). If

these are mutated, the T-DNA is not integrated into the host's nuclear genome. Fusion proteins of the amino acid sequence of these NLSs and glucuronidase are transported successfully into the nucleus. That *virE2* is also active *in planta* is shown by the fact that agrobacteria with defective VirE2 proteins regain their lost pathogenicity when they infect transgenic plants that express *virE2*.

T-DNA integration into the host's genome follows the principle of illegitimate recombination, and is achieved mainly with the host's own enzymes. However, the VirD2 proteins have a special function. It has been shown that the 5' end of the DNA that gets attached to the VirD2 protein is much better conserved than the 3' end without an attached protein. This leads to the conclusion that the 5' end is first attached, via the VirD2 protein, to single-stranded breaks in the host's DNA, giving a starting product that the DNA repair mechanisms can work on. This integration is mostly non-specific, although some short stretches of 5–10 homologous bases can determine the integration site. In any case, the T-DNA is integrated preferentially into transcriptionally active areas of the chromatin. However, the bacterial genes are activated by their own promoters, which are also transferred.

Figure 7.13. Principle of mobilization of the T-DNA from the Ti plasmid. For explanation *see* Box **7.5**. (According Koukolikova-Nicola *et al.* 1987.)

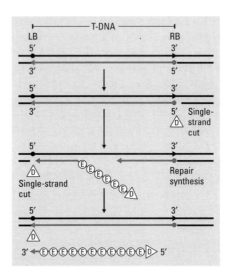

are used and, moreover, cells have nuclei. In fact there are extensive homologies between the *vir* region and some of the *tra* genes that control conjugation. Thus it seems likely that the T-DNA transfer evolved from an ancestor mechanism which had a completely different function.

7.2.3 Agrobacteria as gene vectors

The natural gene transfer mechanisms described in the previous section can be used very successfully to introduce foreign genes into plants (Figure **7.14**). One prerequisite is that the *onc genes* (*tms1*, *tms2*, and *tmr*) are shut off. Plasmids lacking these genes are said to be 'disarmed'. Any desired foreign gene can be introduced between the left and right borders of the T-DNA. In general, as well as the desired gene (suitably equipped with promoter and terminator sequences that the recipient plant will recognize), a marker

Figure 7.14 Derivatives of the Ti plasmid as binary gene vectors. T-DNA, transferred DNA; *vir, vir* region; P1, P2, promoters that initiate expression in the plant; T1, T2, terminators that end transcription in the plant.

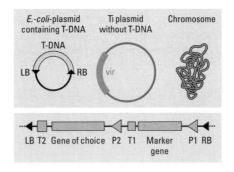

gene will be transferred with similar regulatory sequences. This marker gene might code for resistance to an antibiotic (e.g. neomycin phosphate transferase, NPT II), a herbicide (e.g. BASTA), or a histochemically provable phenotype (e.g. glucuronidase(GUS)). Many different derivatives of the Ti plasmid have been developed for just this purpose. The binary vector systems (e.g. pBIN) have been most success-ful. They share the functions of the Ti plasmid between two plasmids: the larger of the two has the *vir* region; the smaller has the left and right borders flanking the foreign gene. With smaller plasmids only one border sequence is required, because in a small circular DNA this sequence can serve as both the left *and* right border.

Separating the plasmids has the advantage that the for-eign gene can more easily be manipulated in a smaller, more manageable plasmid which can be cultured in *Escherichia coli*. This also makes it easier to select useful restriction enzyme sites.

- Transformation: a DNA sequence stored in *E. coli* can be excised and used to transform agrobacteria. If the bacte-ria contain *vir* plasmids, target plants can be inoculated directly. Naturally, if the transformation works, the only plant cells to be transformed will be the minority into which foreign DNA is transferred. These will have to be selected from the majority of untransformed cells.
- Selection: using markers (such as herbicide or anti-biotic resistance genes) makes selection of trans-formants possible.
- Regeneration: to regenerate the transformed cells that have been selected, a regeneration protocol is needed. The capability of a target plant to be regenerated depends primarily on its genotype. Although many plant species are easily regenerated (especially members of the Solanaceae), a great deal of experimentation has been required to regenerate transgenic plants from other, more recalcitrant, species. However, the number of plant species that can be regenerated routinely grows yearly, so there is little basis for the assumption that there are some plant species that cannot be regenerated at all (and so cannot be transformed). The *Agrobacterium* system is mainly used to transform dicotyledonous plants. Monocotyledonous plants take up the T-DNA, but stable transformants are rare. Interestingly, in recent experi-ments *Agrobacterium* has even been used to introduce genes into yeast cells, so it seems that, in principle, the cell wall and cell membrane are not barriers against transformation by *Agrobacterium*.

7.3 **Rhizobia**

Rhizobia have received much attention from scientists because they can fix atmospheric nitrogen while living in nodules in the roots of legumes (Figure **7.15a,b**). In this way they can supply their host plants with substances that normally need to be supplied by fertilizers.

Initially, expectations for exploiting the symbiosis between plants and rhizobia in agriculture were high. Scientists even dreamed of genetically modifying wheat so that it, too, could use nitrogen from the atmosphere. However, these expectations were later tempered by the realization that the rhizobium symbiosis was much more complex than had at first been thought. Many factors from both the bacterium and the plant need to match before the two can live together. As a result of this, it comes as no surprise that very few plant species are able to fix nitrogen.

As this complexity was discovered, the apparent complication of symbiosis lessened. We now understand the process much better because we understand the interactions between individual factors.

7.3.1 **What can be seen?**

A small selection of bacterial genera (*Rhizobium*, *Bradyrhizobium*, and *Azorhizobium*) are grouped into the Rhizobiaceae family because of their ability to fix nitrogen. In other characteristics they differ so much that they cannot be seen as a monophyletic group. Their host plants are legumes (Fabales) (Caesalpiniaceae, Mimosaceae, and Fabaceae/Papilionaceae). Nevertheless,

Figure 7.15 Root nodules. (a) Nodules on the roots of pea; (b) schematic of infection of lucerne roots. e, Endodermis; i, infection tube (red); r, cortex cells; rh, rhizodermis; wh, root hair; xy, xylem. (From Nultsch, W. *Allgemeine Botanik*, Thieme, Stuttgart, (1996).)

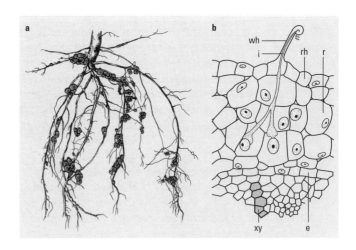

there is an exception to this rule (*Parasponia*, from the Ulmaceae). Associations between rhizobia and legumes are highly specific. Each bacterium, or bacterial strain, is specific to a certain plant host. The way that the bacterium infects the host is determined by the host. Different species of bacteria can gain entry into the host plant using similar mechanisms. For example, they might use an infection thread, or enter by 'crack entry'—splitting the middle lamella between the plant's cells. Also, the type and number of root nodules that develop is determined by the host.

A typical root nodule is shown in Figure **7.16a–c**. It is a specialist organ consisting of a meristem and a network of veins (the xylem and phloem) to supply nourishment to a

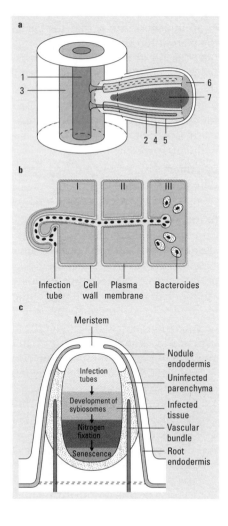

Figure 7.16 Structure and development of root nodules. (**a**) Structure of a root nodule: 1, root vascular bundle; 2, nodule vascular bundles; 3, root endodermis; 4, nodule endodermis; 5, nodule vascular endodermis; 6, nodule meristem; 7, central infection region. (**b**) Schematic diagram of events during infection via the infection thread. I–III, Root cells that are penetrated one after the other. Black lines, plasmalemma; red lines, cell wall. (**c**) Diagram of the different zones during differentiation of a root nodule. (According to Brewin 1991.)

central core where the bacteria replicate and fix nitrogen. The central tissue is surrounded by an endodermis and, as in the root, this is responsible for rigorously excluding the surrounding environment. An excess of a pigment called leghaemoglobin colours the central region red. This leghaemoglobin exists to reduce the amount of oxygen in the cells of the nodule. This is important because bacterial nitrogenase is very sensitive to atmospheric oxygen.

The way that the bacteria force their way into the roots via an infection thread is shown in Figure **7.16b**. The bacteria stop the plants growing root hairs further out into the soil, and change the destinations of membrane vesicles intended for use in the synthesis of the cell wall. The root hair bends back on itself, and the cell walls grow inwardly, pushing through the inside of the parent cell and then on through neighbouring cells. This leads to the establishment of an infection thread that the bacteria can use to invade the plant intercellularly.

At the tip of the infection thread, cell wall synthesis is reduced and the bacteria can force their way through at this point into a vesicle that buds from the plasma membrane. From here, single bacteria are taken up individually by the plant by a process analogous to phagocytosis. The plasma membrane's composition is altered as it acquires nodulation-specific proteins.

Definition The newly formed specialized membrane is called the *peribacteroid membrane*.

The bacteroids divide and differentiate into nitrogen-fixing organelles. Simultaneously, the plant reacts by expressing nodulation-specific genes for leghaemoglobin, glutamine synthase, and uricase, amongst others.

Before the bacteria reach their final destination, nodule production has already been initiated. Contact between the root hairs and the bacteria is sufficient to initiate mitotic activity in a remote tissue (the root cortex). A new meristem emerges that becomes the primordial nodule, the structure of which is shown in Figure **7.16a**. The number and location of root nodules is determined by the host.

The bacteria can force their way into the nodule primordia in one of two ways:
1. By infecting the meristem. This limits meristematic activity, and these nodules are said to be 'determinate'.
2. The bacteria grow inside the infection thread right through the meristem and infect tissue behind it. In this way the meristem remains active, and the nodules are able to grow without limits, up to a certain age. These are 'indeterminate' nodules.

Soybean and *Phaseolus* spp. produce determinate nodules, whereas lucerne, pea, and broad beans produce
indeterminate nodules. The type of nodule produced is
determined by the host. During the formation of indeterminate nodules, sections of polar tissue are constructed.
Figure **7.16c** shows schematically how they develop.
Moving from the outside of the root towards its centre,
the following tissues form: (1) the meristematic zone,
(2) the zone of the infection thread, (3) the zone where
symbiosis develops, (4) the nitrogen fixation zone, and
finally (5) the senescence zone.

The relationship between rhizobia and their host can be
viewed as an unstable balance between parasitism, symbiosis, and saprophytism. Whereas at first the bacteria
seem to behave as pathogens, evading the host's defence
mechanisms as they enter it, later they behave as symbionts, and finally live as saprophytes on dying host cells.
Their complex way of life is regulated by a myriad of host
and bacterial genes.

Definition In general, nodule-specific genes in the bacteria
are called nod genes, whereas the plant-specific genes are
called *nodulins.*

7.3.2 Genetic analysis

How do the partners in this symbiotic relationship recognize one another? How can they regulate each other's
genes for differentiation so that a functional organ develops? How do the rhizobia overcome the host's defences
against invading pathogens? Using molecular genetic
analysis, all of these questions have been answered with
varying degrees of completeness during the past 5–10
years. However, many questions remain. Most of the
advances have been made on the bacterial side, leaving
many unanswered questions about the participation of
host factors. This is mainly due to the current state
of the art of the experimental techniques involved
(Box **7.6**).

The *nod/nol* genes have been localized to large megaplas-
mids, 1.2–1.5 Mb in size (sym plasmids), and are organized
as operons. Examples of these in different rhizobial species
are shown in Figure **7.17**. The nod genes are organized hier-
archically. Some, such as *nod*D, have regulatory functions,
others are structural genes. Some genes are common to all
rhizobia (*nod*ABC), whereas others are specific to certain
host–symbiont complexes.

Box 7.6 Using transposon mutagenesis to identify *nod* genes

Transposon mutagenesis was a key method used to identify the bacterial *nod* genes. The ability of transposons to 'jump' non-specifically (but in a statistically definable way) into host genes and inactivate them made the analysis of nod genes possible. Transposon mutagenesis experiments produced some bacterial clones that either could not induce nodulation, were missing single nodulation functions, or had an altered host range. Since the transposons marked the genes they inactivated, scientists could select just the genes they were interested in for sequencing. By retransforming these clones with a wild-type gene, the gene's function could be proved.

Using these and other analyses, many rhizobial genes (called *nod*, plus an extra letter— see Figure **7.17**) have been identified. The number of these genes has now outstripped the number of letters in the alphabet, so new ones are now designated *nol* plus a letter.

7.3.3 Cross-talk between symbiont and host

To understand how the *nod* genes work, the way the rhizobium and host communicate with each other needs to be known. The processes involved are very similar to the interactions between agrobacteria and plants (section 7.2.2), or those between bacterial pathogens and their hosts. When a plant is wounded or stressed, it secretes many complex organic compounds, including flavones, flavonoids, flavanones, isoflavone, and chalcone (Figure **7.19**) that participate in general defence mechanisms. Rhizobia react chemotactically to these substances, specifically activating their *nod*-specific genes (Figure **7.18**). The specificity of the flavonoids is due to various different chemical substitutions. Rhizobia react very differently to single, or to patterns of, flavonoids, activating different *nod* genes accordingly (Figure **7.19**). At the centre of the signal transduction chain are the *nodD* gene products. They can bind flavonoids to their C-terminal ends. In this way they are activated as transcription factors (Figure **7.18**). The activated NodD protein binds to special structures in other *nod* gene promoters containing a special recognition sequence. These structures are called *nod*-Boxes and are 47 bp long.

Two groups of *nod* genes are activated: common *nod* genes and host-specific *nod* genes. The *nod* genes determine how the bacteria respond to the host plant.

Definition The signalling substances in the signal transduction chain from rhizobia to plant are in fact modified chito-oligosaccharides, called *nod* factors.

The *nod* factors are passed to the plant from the bacteria. Once inside the plant, they promote meristematic activity

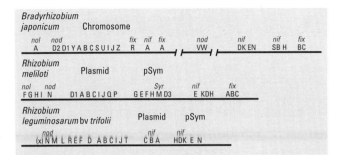

Figure 7.17 Parts of the genetic maps of different rhizobia plasmids which effect the symbiosis. *NodA–D* occur in all rhizobia. Other *nod* genes are species-specific. (According to Triplett and Sadowsky 1992.)

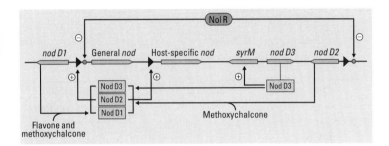

Figure 7.18 Regulation of the *nod* genes by the *nodD* and *nolR* gene products. (According to Dénarié *et al.* 1992.)

in tissues remote from the site where they were produced. This activity leads to the production of nodules. The signalling component itself is built of at least four β-1,4 linked D-glucosamine molecules that can be modified at three sites (R1–R3). These modifications have implications for host specificity. R1 can be acetylated, R2 can carry a sulphate group, and R3 can carry fatty acids of differing degrees of saturation. The chito-oligosaccharides can induce meristem formation at concentrations as low as 10^{-9} M. This means that they are active at lower concentrations than any of the plant hormones known to date. Root hairs can be seen to flex at just 10^{-12} M. The acyl group at R3 seems to aid uptake of *nod* factors across the plant's plasma membrane. As evidence for this, deacylated compounds have no effect if they are supplied externally, unless they are shot inside plant cells using a particle gun (Box **7.7**).

The synthesis of *nod* factors alone is necessary and sufficient to induce root nodule formation. The bacteria themselves do not have to actually further infect the plant, meaning that it is possible for uninfected nodules, which cannot fix nitrogen, to form on the roots. How *nod* factors induce meristem formation and change the plant's morphogenetic scheme is not clear. There are, however, increasing signs that chito-oligosaccharides are not limited

Figure 7.19 Spectrum of effects of plant flavonoids from rhizobial species. NodD induction has been estimated after addition of flavonoids. Rt, *Rhizobium trifolii*; Rl, *Rhizobium leguminosarum*; Rm, *Rhizobium meliloti*; NGR, *Rhizobium* spp. NGR 234; Bj, *Bradyrhizobium japonicum*; n, not determined. (According to Dénarié *et al.* 1992.)

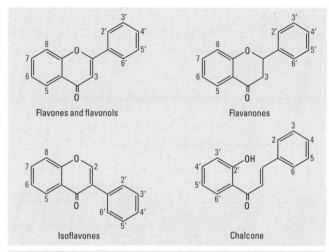

Flavones and flavonols — Flavanones — Isoflavones — Chalcone

Chemical	Position of substitution						NodD-protein					
	3	5	7	3′	4′	5′	Rt	Rl			NGR	Bj
Flavones												
Luteolin		OH	OH	OH	OH		++	++	++	−	++	−
Apigenin		OH	OH		OH		++	++	+	−	++	+
Chrysoeriol		OH	OH	OCH₃	OH		n	n	++	−	n	n
Chrysin		OH	OH				+	−	−	−	++	−
Flavonols												
Myricetin	OH	OH	OH	OH	OH	OH	−	−	−	−	+	n
Quercetin	OH	OH	OH	OH	OH		−	−	−	−	++	−
Camphorol	OH	OH	OH		OH		−	−	−	−	++	+
Flavanones												
Eriodictyol		OH	OH	OH	OH		+	++	+	−	n	−
Naringenin		OH	OH		OH		++	++	−	−	++	−
Hesperetin		OH	OH	OH	OCH₃		−	++	−	−	++	n
Isoflavone												
Genistein		OH	OH		OH		−	−	−	−		
Daidzein			OH		OH		−	−	−	−		
Chalcone												
4,4′Dihydroxy-2′-methoxychalcone					OH		n	n	++	++		

only to playing a part in symbiosis, they may also be important for basic plant development. Also, in non-host plants, such as carrots, *nod* factors could help to overcome problems during the development of mutants in somatic embryogenesis.

The host plant must be susceptible to the *nod* factors. Several plant genes that are involved have been mapped using both classical methods and molecular tools. This has revealed that, as with pathogenesis, there is a gene-for-gene relationship in symbiosis. In 1956 Flor stated his 'gene-for-gene hypothesis', in which he said that genes

Box 7.7 Functions of the nod genes

One function of the *nod* genes is to participate in producing the signalling substances (Figure **7.20**). *NodM* codes for glucosamine phosphate synthetase. *NodC* takes part in synthesizing the chitin oligomer backbone and shows homologies with the chitin synthase of yeasts. *NodB* deacylates the first glucosamine group so that *NodA* can acylate the freed bond position. *NodH* and *NodPQ* add a sulphate group at position R2, and *NodL* acetylates R1. NodF has homologies with acyl carrier pro-

teins, and *NodE* is responsible for producing highly unsaturated fatty acids. These examples demonstrate the scope for genetic engineering to change the *nod* factors.

Not all *nod* genes, though, participate in the synthesis of *nod* factors (chito-oligosaccharides). For example, *NodO* encodes a Ca^{2+} binding protein. It has homologies with haemolysin from *E. coli*, and forms ion channels across the bacterial membrane for monovalent cation exchange.

that are important for the co-existence of two organisms will evolve together in pairs. For example, these could be genes for virulence and susceptibility, or for virulence and resistance (Box **7.8**).

It is possible that lectins may be important for recognition. A lectin gene from pea was genetically engineered into *Trifolium pratense* L. This transformed white clover was found to be susceptible to pea-specific rhizobia. Pea roots are rich in lectins, which, at the cellular level, are located on the exterior of the plasma membrane. The *nod* factors are effective ligands for lectins.

Plant genes are not only responsible for governing which rhizobia can infect and produce root nodules. They also regulate the location and number of nodules formed. Soybean lines exhibiting either super-nodulation or no

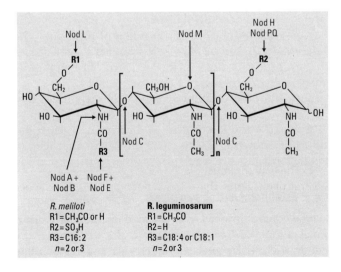

Figure 7.20 Chemical structure of *nod* factors of *Rhizobium meliloti* and *Rhizobium leguminosarum*. The species-specific structure is determined by the constituent parts R1, R2, and R3. (For an explanation *see* Box **7.7**.)

Box 7.8 Specificity of rhizobium–legume symbioses

The interaction between Bradyrhizobium japonicum strain USDA 123 and soybean is a good example of a rhizobium–legume symbiosis. Several soybean strains have been isolated that specifically hinder nodulation by B. japonicum. A single dominant plant gene is responsible for this, but a mutation in the nolA gene of the bacterium can overcome this resistance.

In a similar fashion, most Rhizobium legu-minosarum strains cannot induce nodulation in Afghan pea plants. Again, this is due to the effect of a single gene. However, the R. leguminosarum strain TOM overcomes this resistance by carrying the nodX gene.

These genetic experiments hint at the existence of a frequently used recognition mechanism, which acts according to the 'lock and key' principle of gene products from bacteria and plants.

---

nodulation at all have been isolated. These two lines have been root–stem grafted to each other experimentally, and the results were interpreted as meaning that a diffusion signal is responsible for inhibiting production of new nodules near existing ones. According to this model, the super-nodulation line nts382 is unable to produce this sort of inhibiting signal. Currently, much research is being focused on isolating the gene responsible for the signal (*see* Chapter 3).

Some nodulin genes were originally believed to be expressed only in the nodules. However, it has now been shown that identical or similar genes are also expressed in the plant for completely different purposes. This is also true for the leghaemoglobin gene. Initially, the apparent uniqueness of this gene in plants led to speculation that a horizontal gene transfer from the animal world might have occurred. However, it is now known that many plant species possess haemoglobin-like genes, which regulate the concentration of oxygen in different processes. The gene *nodulin* 26 can be viewed in a similar way—it codes for a transport protein in the peribacterial membrane, which is also a variation of a tonoplast protein.

These and other examples support the hypothesis that the plant–rhizobia symbiosis is the product of a complex on-going co-evolution of plant and bacterial genes. It has still to be shown how the common signals that lead to nodule formation are regulated.

7.3.4 Establishment of symbiosis

Several conditions need to be satisfied for successful nitrogen fixation. An important factor is the structure of the nodule itself (Figure **7.16**). The tightly packed surrounding cells, enveloped in an endodermis, help to isolate the bacteria from atmospheric oxygen. To finish the job, leghaemoglobin is synthesized in large amounts, accounting for up to 25 per cent of the total cytoplasmic protein. The bacteria activate their *nif* genes, which code for nitrogenase. To detoxify and utilize the NH_4^+ produced by the bacteria, glutamine synthetase is accumulated. Sometimes, the molecularly fixed nitrogen is transported with the aid of uricase as allantoin/allantoic acid within the plant.

The peribacterial membrane differentiates and takes on a key role, transporting specific substances. Between the bacterial and peribacterial membranes, the glycocalyx is deposited. To supply the bacteria with sources of carbon dioxide (chiefly malate and succinate), there is a dicarboxylate carrier in the peribacterial membrane. Between the bacterial and peribacterial membranes neither cell wall substances from the plant nor exopolysaccharides from the bacteria are deposited. This space is reserved for a dedicated transfer system for the substrates and products of symbiotic nitrogen fixation.

7.4 Bibliography

Viruses and viroids

Agrios, G. N. (1997). *Plant pathology*. Academic Press, San Diego.
  Basic textbook.
Bos, L. (1978). *Symptoms of virus diseases in plants*. Centre for Agricultural Publishing and Documentation, Wageningen.
  Good description of the variety of virus symptoms.
Citovsky, V. and Zambryski, P. (1993). Transport of nucleic acids through membrane channels: Snaking through small holes. *Annual Review of Microbiology*, 47, 167–97.
  Review of the latest knowledge about transport of viral and agrobacterial nucleic acids in plants.
Collmer, C. W. and Howell, S. H. (1992). Role of satellite RNA in the expression of symptoms caused by plant viruses. *Annual Review of Phytopathology*, 30, 419–42.
  Review.
Gilbertson, R. and Lucas, W. J. (1996). How do viruses traffic on the 'vascular highway'? *Trends in Plant Science*, 1, 260–8.
Matthews, R. E. F. (1991). *Plant virology*. Academic Press, San Diego.
Moffat, A. S. (1994). Mapping the sequence of disease resistance. *Science*, 265, 1804–5.
Palukaitis, P., Roossinck, M., Dietzgen, R. G., and Francki, R. I. B. (1992). Cucumber mosaic virus. *Advances in Virus Research*, 41, 281–347.
  Detailed description.

Agrobacteria

Baron, C. and Zambryski, P. C. (1995). The plant response in pathogenesis, symbiosis, and wounding: variations on a common theme? *Annual Review of Genetics*, 29, 107–29.

Binns, A. N. and Thomashow, M. F. (1988). Cell biology of *Agrobacterium* infection and transformation of plants. *Annual Review of Microbiology*, 42, 575–606.

Hooykaas, P. J. J. and Schilperoort, R. A. (1984). The molecular genetics of crown gall tumorgenesis. *Advances in Genetics*, 22, 209–83.
Detailed description of classical and molecular genetic knowledge of tumogenesis.

Klee, H., Horsch, R., and Rogers, S. (1987). Agrobacterium-mediated plant transformation and its further applications to plant biology. *Annual Review of Plant Physiology*, 38, 467–86.

Koukolikova-Nicola, Z., Albright, L., and Hohn, B. (1987). The mechanism of T-DNA transfer from *Agrobacterium tumefaciens* to the plant cell. In *Plant DNA infectious agents*, (ed. T. Hohn, and J. Schell), pp. 109–48. Springer, Wien.
Early article.

Lippincott, J. A. and Lippincott, B. B. (1975). *The genus Agrobacterium* and plant tumorigenesis. *Annual Review of Microbiology*, 29, 377–405.

Zambryski, P. C. (1992). Chronicles from the Agrobacterium–plant cell DNA transfer story. *Annual Review of Plant Physiology and Plant Molecular Biology*, 43, 465–90.
Extension of Koukolaikova-Nicola's work.

Rhizobia

Brewin, N. J. (1991). Development of the legume root nodule. *Annual Review of Cell Biology*, 7, 191–226.
Basic review.

Dénarié, J. and Cullimore, J. (1993). Lipo-oligosaccharide nodulation factors: a minireview. New class of signaling molecules mediating recognition and morphogenesis. Cell, 74, 951–4.
Good description of the interaction between host and rhizobia.

Dénarié, J. and Debellé, F. (1996). Rhizobium lipo-chitooligosaccharide nodulation factors: signaling molecules mediating recognition and morphogenesis. *Annual Review of Biochemistry*, 65, 503–35.

Dénarié, J., Debellé, F., and Rosenberg, C. (1992). Signaling and host range variation in nodulation. *Annual Review of Microbiology*, 46, 497–531.

Hunt, S. and Layzell, D. B. (1993). Gas exchange of legume nodules and the regulation of nitrogenase activity. Annual *Review of Plant Physiology and Plant Molecular Biology*, 44, 483–511.

Leigh, J. A. and Coplin, D. L. (1992). Exopolysaccharides in plant–bacterial interactions. *Annual Review of Microbiology*, 46, 307–46.
Good description of the ability of rhizobia to overcome host defences.

Nultsch, W. (1996). *Allgemeine Botanik*. Thieme, Stuttgart.

Spaink, H. P. (1995). The molecular basis of infection and nodulation by rhizobia: the ins and outs of sympathogenesis. *Annual Review of Phytopathology*, 33, 345–68.

Triplett, E. W. and Sadowsky M. J. (1992). Genetics of competition for nodulation of legumes. *Annual Review of Microbiology*, 46, 399–428. Description of the genetic basis of host–rhizobia interactions.

Glossary

Abaxial The part of a plant that faces away from the shoot, such as the underside of a leaf.

Action spectrum Measurement of the effectiveness of photons for a reaction parameter with respect to their wavelength. Action spectra allow the determination of the absorption peak of a photoreceptor which participates in a physiological reaction.

Adaxial The part of a plant that faces towards the shoot, such as the upper side of a leaf.

Affinity chromatography A chromatographic method in which the specificity of ligand binding to a protein is used to purify that protein.

Albino A mutation in many genes that autonomously hinders the accumulation of pigments in cells.

Allele A sequence variation of a stretch of DNA (a gene); mutant alleles are usually recessive to their dominant wild-type allele. Molecular alleles can be co-dominant, recessive, or dominant.

Alloploidy Multiple sets of chromosomes that have emerged by hybridization of different species (for example, *Triticum*).

Androecium Collective name for the male parts of a flower.

Anther Pollen-bearing part of the stamen.

Antibody staining Method showing the distribution of gene products in a tissue using indirect immunofluorescence or immunostaining with specific antibodies.

Anticline An angle perpendicular to the surface.

Antisense RNA RNA with a sequence complementary to the target RNA. Antisense RNAs produced by *in vitro* transcription are used for *in situ* hybridization. Expression of antisense genes in plants is used to lower gene activity.

Apical/basal Main axis of polarity within the plant. Can already be seen in the zygote.

Apical dominance Suppression of lateral shoot formation by the shoot meristem.

Apomixis The asexual formation of fertile seeds; diploid embryos can emerge from diploid embryo sacs because of defective meiosis in the embryo sac mother cell or a later-formed diploid embryo sac replaces the haploid original.

Archespore Tissue inside the pollen sac.

Autopolyploidy Multiple sets of chromosomes which emerged by self-replication.

Autoregulation A gene product controls its own expression.

Cadastral genes Early acting regulatory genes of flower development which influence the local expression of homeotic genes, which are expressed later.

Cadherin Transmembrane glycoprotein of the plasma membrane which provides cell–cell adhesion in animal tissues. It is Ca^{2+} dependent.

Calmodulin A Ca^{2+}-binding protein which can activate specific protein kinases if it is complexed with Ca^{2+}.

CAPS DNA polymorphism which can be shown after PCR amplification with sequence-specific primers and digestion of the amplified products with restriction enzymes.

Carpels Female flower organs.

Casparian strip Endodermal barrier against substance transport in the plant.

Cell autonomy Differentiation of a cell depending on its genotype.

Cell file Regular arrangement of cells along an axis, for instance in the root. They are partially of clonal origin.

Chalaza The basal area of the ovule.

Chimera An individual composed of cells with different genotypes. It is derived from different zygotes, in comparison to a genetic mosaic.

Chromosome walking Term for the identification and arrangement of overlapping DNA clones during map-based cloning.

Clonal analysis Examining the developmental fate of a cell by marking it genetically.

Clone Genetically identical offspring of a single cell.

Coat protein The protein surrounding and protecting the nucleic acid of a virus.

Compartment Intracellular reaction space, separated from the rest of the cell by a membrane.

Competence The ability of a cell, organ, or entire organism to achieve a particular developmental step.

Complementation Functional proof of the successful cloning of a gene by the suppression of a mutant phenotype of an allele by transformation.

Complementation test A test to determine whether two mutants carry mutations in the same or different genes. If there is no complementation, the mutations are within alleles.

Complexity (of a sequence) The complete lengths (in nucleotides) of sequences present in a DNA or RNA population. This means that every sequence that is present more than once is only counted once.

Connective Tissue between the two thecae of an anther.

Corpus Inner layer (L3) of the shoot apical meristem.

Cosmid Vector for cloning up to 40 kb DNA fragments in bacteria.

Co-suppression Interdependent inhibition of expression of multiple gene copies.

Cotyledon Leaf formed during embryogenesis. It often differentiates into a storage organ.

Cross-protection Protection mechanism against virus diseases by the introduction of a mild virus or its genes.

Cytoplasmic pollen sterility Mitochondrial gene defect which leads to disruptions in pollen development; the plants are male-sterile.

Determination The commitment of a cell to a certain, almost irreversible differentiation pathway.

Developmental gene A gene that affects development. Its effects can be noticed if a mutant phenotype arises in its absence.

Differential display Very effective technique for analysis of phase, tissue, or organ-specific RNA populations using reverse transcription and PCR (RT-PCR).

Differentiation The process by which cells, tissues, and organs become different in structure and function.

Differentiation zone Area of the root above the zone of elongation.

Distance between gene and marker, two genes, or two markers is measured genetically as the relative distance estimated by recombination frequency in centiMorgans (1 cM = 1 per cent of recombination); the actual physical distance is measured in kilobases (kb).

DNA polymorphism A DNA sequence difference between homologous parts of the genomes of two different individuals which can be shown experimentally.

DNA reassociation kinetics Measurement of the kinetics of renaturation of double-stranded DNA. Repetitive DNA sequences reassociate more quickly than non-repetitive sequences. In this way the degree of repetition of DNA sequences and their abundance throughout a genome can be measured.

Dorsoventral The axis between the upper and lower sides of a leaf.

Ecotype Geographically restricted population of a species.

Ectopic expression Expression of a gene outside its usual domain, for instance by fusion of a cDNA to a constitutive promoter.

Embryo sac The female gametophyte of seed plants.

Endodermis Layer of ground tissue forming a sheath around the vascular region; is the innermost layer of the cortex in roots and stems of seed plants.

Endoreplication DNA replication without nuclear division.

Endosperm Triploid nutritive tissue for an embryo. Arises from the fertilization of the diploid central cell.

Endothecium Second outer wall surrounding the pollen sac.

Enhancer *Cis*-regulatory sequence element of transcription; its position and polarity varies in relation to the starting point of transcription. It can also enhance transcription from a distance.

Epistasis The phenotype of a mutant inhibits the expression of the mutant phenotype of a different gene of another mutant; this hints that there is a regulatory interaction between the two genes.

Evocation Transition from the vegetative shoot meristem to the generative inflorescence meristem.

Exine The outer layer of a pollen grain.

Family Offspring of a selfed plant.

Fasciation A split in the shoot which occurs, for instance, when the meristem enlarges.

Fibronectins Extracellular glycoproteins which help the cells of vertebrates to adhere to the extracellular matrix.

Flower meristem identity genes Control genes for specific characteristics of flower meristems.

Footprinting Method that helps in identification of the binding site of a DNA-binding protein.

Fucanes Fucose-containing heteromeric cell wall polysaccharides of brown algae.

Funiculus The stalk of the ovule connecting it to the placenta.

Gametophyte Gamete-forming generation of an organism.

Gene amplification Multiplication of a gene so that two or more copies, usually at the same site, are present in the genome.

Gene fusion The joining, by recombinant DNA techniques, of two or more genes that encode different products.

Generation Part of the ontogeny of an organism that begins with a particular germ type and which, after mitoses, ends with a different germ type (gametophyte, sporophyte).

Generation change Regular change of a generation by different types of propagation.

Gene tagging Marking a gene using insertional mutagenesis. This process makes it possible to clone by inverse PCR or plasmid rescue.

Genetic mosaic An individual composed of cells with different genotypes. Unlike chimeras, they develop after a change in the genotype of single cells during development.

Genome equivalent Statistical measurement for the completeness of a genomic DNA library; if there are five genome equivalents present, 99 per cent of the genome will occur at least once in the DNA clones.

Germ line Cells of the generative cell line, which in animals are separated at a very early stage from the somatic cells.

Gynoecium Collective term for the carpels of the flower.

Helper virus A virus that aids the replication of another defective virus infecting the same cell.

Heterotrimeric G proteins Parts of the signal transduction chain composed of three different subunits (α-, β-, and γ-); only the α-subunit binds GTP or GDP and possesses an endogenous GTPase activity.

Homeobox A conserved DNA sequence of about 180 bp which has a DNA-binding motif in its deduced amino acid sequence. It was first discovered in the homeotic genes of *Drosophila*.

Homeostasis Maintenance of a stable internal state of a cell or an entire organism.

Homeotic gene Genes with functions for formation of specific organ identities.

Homology Describes in evolutionary biology the identical origin of phenotypes, independently of the original functions of each phenotype. In molecular biology this term is often used as an incorrect synonym for sequence identity.

Hybrid bleaching A term from classical plant genetics which describes the emergence of chlorophyll deficiencies in interspecific crossings (species hybrids).

Hybridization In molecular biology, the pairing of complementary RNA–RNA, RNA–DNA or DNA–DNA strands to make a hybrid. In genetics, it describes either the crossing of different species or two inbred lines of a single species.

Hypersensitive site Short area of the chromatin where the DNA is extremely accessible to restriction by nucleases such as DNase I.

Hypophysis The uppermost cell of suspensor.

Idioblast A cell in a tissue that markedly differs in form or size from other cells in the same tissue.

Immunoblotting Qualitative and semi-quantitative visualization of proteins. After separation by gel electrophoresis proteins are transferred to a membrane and then detected by incubation with specific antibodies.

Immunofluorescence/immune staining The visualization of particular proteins within a tissue by the use of fluorescence-tagged antibodies of enzyme-linked secondary antibodies which bind specifically to the first antibody.

Inflorescence Arrangement of flowers on a floral axis.

Initial Stem cell in meristems. It forms cell lineages.

Inositol-1,4,5-triphosphate A chemical produced from membrane lipids by the action of phospholipase C. It forms part of the signal transduction chain.

Insertional mutagenesis The alteration of a gene sequence by inserting foreign DNA sequences into it. For example, by transformation using T-DNA or by mobilizing transposons.

In situ **hybridization** A technique used to locate within a cell transcripts that are complementary to a specific labelled probe.

Integrin Receptor proteins at the cell surface with which the cells bind to the extracellular matrix.

Integument Outer cell layers enveloping the nucellus of the angiosperm ovule.

Intercalar Non-meristematic cell division.

Intine Inner wall of a pollen grain.

Inverse PCR Amplification of DNA fragments using primers which point outwards. Requires the DNA to be circular.

Kozak's rule According to M. Kozak, eukaryotic ribosomes scan mRNA from its 5′ end until they happen on the first AUG (start) codon in a known sequence context. From here they commence translation. Further AUG codons in the 3′ direction are not used for translation initiation.

Leader Non-translatable part of an mRNA upstream of the start codon.

Lectin A protein that binds sugars specifically.

Leghaemoglobin Haemoglobin of leguminous plants.

Leucine zipper A structural motif of a class of transcription factors, used as a dimerization domain.

Linkage An expression meaning that two genes are on the same chromosome, and are hence inherited together.

Locus In classical genetics, the position of a gene on a chromosome as it can be described by genetic mapping. These days the term 'locus' is mainly used to describe a chromosomal region that contains one or several consecutive genes.

Locus control region DNA sequence which has a higher-level control function at the level of chromatin for every gene within a gene group. The term was first used for the β-globin gene locus in humans.

Map-based cloning Cloning of genes by mapping them against molecular markers followed by chromosomal walking.

Marker Differences that can be observed at the phenotypic or molecular levels and which can be used to map a gene.

Megagametophyte Female gametophyte.

Megasporangium Sporangium that produces megaspores.

Megaspores Spores which form the megagametophyte.

Megasporophyl Leaf-like organs that carry the megasporangia.

Microgametophyte Male gametophyte.

Micropyle Opening between the integuments of the ovule through which the pollen tube can penetrate the ovule.

Microsporangium Sporangium that produces microspores.

Microspores Spores which form the microgametophyte.

Microsporophyl Leaf-like organs that carry the microsporangia.

Morphogen Substance that causes the formation of a particular shape.

Morphogenesis Change of form of a developing organism. In plants this is established by local cell activities (division and elongation).

Movement protein A protein that aids movement of viral nucleic acids from cell to cell.

Mutagen Something that increases the mutation frequency. It can be a chemical or ionizing radiation. Other mutagens such as T-DNA and transposons have been used specifically for gene tagging.

Necrosis Symptoms of the death of tissues and organs.

nod **gene** Bacterial gene that induces nodulation in plants.

Nodulin gene Plant gene that is activated specifically for nodule formation.

Nopaline Amino acid derivative specifically formed by plants after they are infected with agrobacteria.

Octopine Amino acid derivative specifically formed by plants after they are infected with agrobacteria.

onc **genes (oncogenes)** Genes responsible for tumour development.

Operon Consecutive genes which are transcribed as a set. They are typical in the genome organization of prokaryotes.

Organ Part of the body that is composed of different tissues.

Ovule The megasporangium of seed-forming plants which is composed of the inner and outer integuments, the nucellus, and the embryo sac.

Pattern The non-random distribution of structures in a body or an individual organ.

PCR amplification Multiplication of DNA sequences using thermostable DNA polymerases and oligonucleotides as primers.

PCR marker A general term for DNA polymorphisms that can be shown by PCR amplification. This term is often used as a synonym for CAPS markers.

Peptide growth factor Extracellular polypeptide with signal characteristics which can cause a cell to grow or proliferate. Examples are fibroblast growth factor (FGF) and the transforming growth factor-β (TGF-β).

Perianth Collective term for petals and sepals.

Periclinal Oriented parallel to the surface.

Periclinal chimeras Plants in which one of the cell layers of the apical meristem differs genetically from the others.

Pericycle The outer cell layer of the vascular bundle which borders the endodermis. Also known as the pericambium. It forms side roots.

Petiole Stalk which attaches the leaf blade to the stem.

Phenotype The external manifestation of a genotype.

Photoaffinity labelling Visualization of proteins using ligands that are activated by light.

Photoinhibition Drastic reduction of the rate of photosynthesis in high light fluxes. Photosystem II, in particular its D1 protein, is affected most.

Photo/scotomorphogenesis Developmental strategies of plants in either light or dark conditions.

Phyllotaxis The mode in which the leaves are arranged on the axis of a shoot.

Physical gene map Arrangement of overlapping cloned genomic DNA fragments.

Phytomere A repeating element along the plant body axis. Each phytomere consists of the node with the attached leaf, and the internode below with the axillary bud at its basis.

Plasmid rescue Cloning in bacteria of, for example, flanking DNA sequences after insertional mutagenesis of plants. The insertional mutagen carries plasmid sequence that allow replication in a bacterial host.

Polarity Physiological or structural differences along an axis.

Polyadenylation Addition of poly(A) to the 3' end of a transcript.

Polyploidy Presence of more than two sets of chromosomes in a cell nucleus.

Polyprotein Primary translation product which is digested with proteases to yield functional subunits.

Primary leaves The first two leaves to form after the cotyledons.

Primordium Morphologically recognizable group of cells in their earliest stage of differentiation which forms a structure (a tissue or organ).

Procambium Primary meristem or meristematic tissue which differentiates into the primary vascular tissue.

Promoter The complete sequence elements near the starting point of transcription. Used by the RNA polymerase and its factors to initiate transcription.

Protoderm The primordial embryo tissue that will later form the epidermis of the plant.

Proto-oncogene The cellular counterpart to a variety of viral genes (oncogenes) that will cause transformed cells to become malignant.

Proximo-distal Longitudinal axis of an organ, for example a leaf, from the base to the tip.

PR proteins Pathogen-related proteins induced by wounding or stress in plants as part of a defence mechanism.

Pseudogene A non-functional gene.

Pulsed field electrophoresis Electrophoretic technique which uses alternating electrical fields (pulses) to separate large DNA fragments.

Quiescent centre Core of the root meristem which contains very slowly dividing cells.

Radial plane The plane that is perpendicular to the apical–basal axis.

RAPD DNA polymorphism which can be recognized by DNA amplification using only a single primer. For this to work, there must be inverted primer targeting sites close together.

Receptor protein kinase Membrane-bound protein kinase with extracellular receptor domain.

Recombination The process which generates a haploid product of meiosis whose genotype is different from either of the two haploid genotypes that constituted the meiotic diploid; basis for segregation analysis and map-based cloning.

Recombination frequency The proportion (or percentage) of recombinant cells or individuals.

Regeneration Regrowth of a lost body part; also the formation of a new plant from callus tissue.

Replication Multiplication of genomic RNA or DNA.

Reporter gene Gene that is not normally present in the host cell. Its activity can be shown by the use of a simple procedure.

Resistance gene Gene that confers resistance against pathogens, chemicals, and other malign influences.

Retinoic acid Derivative of vitamin A. Used by vertebrates for development-specific processes as a local mediator.

RFLP DNA polymorphisms that can be revealed using hybridization probes as different restriction patterns.

Root cap Protective layer around the root tip, composed of a central and a lateral layer which derive from different initials in the root meristem.

Root collet Area of the root that borders the hypocotyl.

Root hair cell One of the two epidermal cell types of roots which form the root hairs.

Satellite RNA RNA that requires a helper virus for its replication. It is not related to the helper virus.

Saturation mutagenesis Induction and recovery of large numbers of mutations in one area of the genome, or in one function, the aim of which is to identify all genes participating in a particular process by examining their mutant alleles.

Sector An area of tissue that differs from the surrounding tissue phenotype and is of a common clonal origin.

Sectored chimera These emerge when single cells in an apical meristem are genetically altered.

Segregating population F2 offspring of a crossing between genetically different individuals.

Segregational analysis Estimation of the distances between a gene and a marker by use of their recombination frequencies.

Single line The offspring of a single individual produced by selfing or crossing.

Spliceosomal introns Nuclear introns; these are removed from the primary transcript by a spliceosome, an RNA–protein complex.

Sporangium Container composed of one or several cells in which the spores develop.

Spores Propagative cells which have emerged asexually by division. One or more spores are formed by division on the inside of a container (a sporangium); spores can be made during mitosis (mitospores) or during meiosis (meiospores).

Sporoderm Wall layer of the pollen grain which is composed of the exine and intine.

Sporophyte Spore-forming generation of an organism.

Sporopollenin Long-chain chemically resistant substances found in pollen grains and fungal spores. Composed of isoprene units.

Stem cell Undifferentiated cells which are committed to use a single (unipotent) differentiation pathway, or several (pluripotent) differentiation pathways. Stem cells divide unequally; after division one daughter cell remains a stem cell while the other cell differentiates.

Suspensor An extension at the base of the embryo that anchors the embryo in the embryo sac.

Sympetal Petals which have grown together at their bases to form a tube.

Symplast Cell population connected by plasmodesmata.

Synteny Condition when the chromosomal arrangement of genes between species is conserved.

Tapetum Inner layer of the pollen sac wall. Necessary for nourishing the archespore within.

TATA box TATA-containing sequence element of promoters which is recognized by RNA polymerase II; it binds the transcription factor TFIID.

Theca A part of the anther.

Theoretical resolution The limit of recombination between a gene and a marker that is observable. Based on the size of the segregating population.

Transgenic plants These have a foreign DNA fragment(s) integrated into their genomes after transformation.

Transient expression Occurs when a foreign gene is not stably integrated into the genome of cell or an organism. The foreign gene is only intermittently active, and is degraded over a period of time.

Transposon Mobile DNA fragment with recognition sequences for the transposase enzyme which it encodes. Endogenous transposons, or those which are integrated by transformation with T-DNA (which are modified and heterologous), are used for gene tagging.

Tunica The L1 and L2 layers of the shoot meristem.

Two-component system Signal recognition and transduction system which was initially identified in prokaryotes. Composed of a histidine sensor kinase with receptor functions and a response regulatory protein which is activated by phosphorylation of a conserved aspartate residue. The response regulator often acts in prokaryotes as a transcription factor.

Vernalization Stimulation of flower development by treatment with low temperature.

vir **genes** Genes that determine the virulence of a pathogen.

Vitronectin An adhesive molecule found in the extracellular matrix of animal cells.

YAC Yeast artificial chromosome. A vector composed of functional elements of a yeast chromosome (centromere and telomere), a selectable marker, and a cloning site. Used to clone DNA fragments which are several hundred base pairs long into yeast cells.

Zone Functional unit of the shoot meristem; the central zone for the renewal of the meristem, the peripheral zone, for formation of the leaf primordia and the secondary meristem, and the ribbed zone for shoot elongation.

Zone of elongation Area of root above the meristem. Cells in this area are actively elongating.

Zygomorph Flowers with only a single plane of symmetry.

Index

Note: Figures and tables are indicated by *italic* page numbers, boxes by **bold numbers**.

ABC model [of flower organ identity genes] 200–2, *202*
abscisic acid
chemical structure *135*
effects *135*, 134, 146–7, 151–2
abutilon mosaic virus 223, *223*
genome structure *227*
Acetabularia alga
blue-light effects 127
cell-differentiation studies 20
reproduction 12
acetosyringone, effects *240*, **241**
actinomorphic flower 199
adhesion mechanisms 30
affinity chromatography 140, 256
agrobacteria 235–43
as gene vectors 242–3
pathological effects on plants 3, 20, 221, 236, 237
transfer of genes into plants 241–2, **241**
Agrobacterium rhizogenes 134
Agrobacterium tumefaciens 3
defensive signals 221
regulation of genes 140, 237–43
Ti plasmid 237, *238*
T-DNA 237, *238*
vir region 237, *238*
tranfer of T-DNA into plant cells 30, **41**, 46, 58, 59, *238*, 239
tumour generation caused by 237
albumin proteins **148**
algae, reproduction 12
alleles, meaning of term **48**, 256
Alleopathic interactions 137

Alloploidy 71, 73, 256
Amaranthus spp. 87, *194*
amygdalin 152
anaphase 16
animal cells
embryogenesis 32
embryonic development 34–5
anthocyanin synthesis, regulation of 81–4
antibiotic-resistance marker gene 59, *61*
Antirrhinum majus [snapdragon] 197
flower organ identity mutants 199, *201*
flower organs 198–9, *198*, *199*
leaf arrangement 198; *198*
transcription factors *81*
transposons 60
AP2 (apetala2) proteins *81*
apical–basal pattern [in seedlings] 167–8
mutants in *Arabidopsis* 168–71
apical dominance 33, 182, 256
apical meristems 35, *35*
see also root ...; shoot meristem
Arabidopsis thaliana [thale cress] 1, **41**
abscisic acid-absent mutants 152
abscisic acid-insensitive mutants 149–50
ain/ctr/ein/eto/etr mutants 141
blue-light effects 127
blue-light receptors 127–8
det/cop mutants 124–7
effects of ethylene 141–4
effects of light/darkness *114*

embryogenesis 163–7
flower-formation features *210*
flower organ identity mutants 199, *201*
flower organs 197–8, *198*, *199*
fusa/fus3 mutants 125, 150
gametophytic mutants 211–12, *213*
gene families 76
genome analysis 71–2, *72*
genome size 52, 70, *71*
genomic DNA in YAC library 57
hy mutants 122–3, 126, 128
leaf arrangement 180, 197, *198*
leaf development 186–7, 197
light induction of seed germination 119, *119*
molecular markers **53**
pattern-formation mutants 167–8
photoperiodic control of flowering *194*, 194–6
phytochrome gene family 121–2, *122*
root-development mutants 177, *178*, 179
shoot development 180, 184–6, 197
transcription factors *81*
transposons 60, *61*
trichome development 187–90, *191*
archaebacteria, cellular composition 7–8
archespore 210, 256
ARE (anaerobic response element) *85*
Artemisia, alleopathic interactions 137

asexual reproduction   10
autonomous gene effects   50–1
autopolyploidy   71, 256
autoregulation [of gene
      expression]   *122*, 256
autospores   12
autotrophy   157
auxin-binding proteins   141
auxins
   chemical structure   *135*
   effects   *135*, 136, 144–5
Avena sativa [oats]   194

bamboo, shoot-elongation rates
      16
basic attributes [of life]   9
Begonia semperflorens   194
Bertholletia excelsa [brazil nut],
      storage proteins   **148**
Beta vulgaris [beet]
   genome size   *102*
   photoperiodic control of
         flowering   *194*
bHLH proteins, transcription
      factors for   *81*
bibliographical databases   4
bilateral symmetry   18
biological clock   130–3
   gene expression driven by
         131–3
   process model for   *133*
biparental transmission
      mechanism   88
blue light, morphogenesis
      affected by   127
blue light receptors   127–9
blue-light-regulated gene
      expression   129–30
Bradyrhizobium japonicum
   effect of flavonoids   *250*
   genetic maps of plasmids
         249
   symbiosis with soybean   ⋅ *252*
Brassica spp.
   B. napus [oilseed rape]
   phytohormones   139
   self-incompatibility   213
   chondriomes   101, *102*
brassinosteroids   *135*, 138
Bryonia dioica [white bryony],
      bending reaction   137
Bryphyllum plants, reproduction
      10

bZIP-type transcription factors
      81, *121*

C4 plants
   bundle sheath and mesophyll
         chloroplasts,
         differential expression
         of plastid genes   100,
         101, *102*
   cell types   22
   photosynthesis   77
CAB genes   76–7
cadastral genes   207, 208, 256
cadherins   29, 256
Caenorhabditis elegans
      [nematode]   1
   genome size   70, *71*
callus cells, differentiation of
      27–8
calmodulin   124, *125*, 256
CAM (crassulacean acid
      metabolism) plants,
      photosynthesis   77
cancers   19
Cannabis sativa [hemp], sex
      change in flowers   136
CAPS (cleaved amplified
      polymorphic stretch)
      53, 256
Casparian strip   177, 256
cauliflower mosaic virus
   35S promoters   61, **64**,
         77–80
   structure   *80*
   genome structure   *227*
cell cycle
   interphase   15–16
   M phase   15, 16
cell differentiation   17
   genes affecting   20–1
   reasons for   20–4
   stimuli affecting   20
cell elongation   16, 176–7
cell lineages   36, 256
cell polarity   23–4, *25*, 33–4
cell-to-cell contact   29–30
cell wall polysaccharides   24,
      30
chalaza   212, 257
chalcone, chemical structure
      250
chalcone synthase   82

chalcone synthetase [gene]
      promoter   120, *121*,
      124, *125*
chemical stimuli, cell
      differentiation affected by
      20
chimeras   50–1, 257
chito-oligosaccharides   138–9,
      249
Chlamydomonas alga
   genomes   90, 101, *102*
   reproduction   12
Chlorella alga, reproduction
      12
chlorophyll, in chloroplast
      biogenesis   156–7
chloroplast development   89,
      89, **89**, 115, 154–7
chondriomes   101–7
   coding capacity   102, *103*
   size   101, *102*
chromatin, organization of
      84–5
chromosomes   8
   in meiosis   10
chromosome walking   54–7,
      257
Chrysanthemum hort.   194
circadian rhythm   130, 195
cis-regulatory sequences   120,
      122
clonal analysis   **182**, 257
   leaf development   186–7
   root development   173–4
   shoot development   182–3
Coffea arabica   194
colour changes, virus-caused
      224
competence, meaning of term
      25, 28, 257
complementation groups   46,
      47, **48**
complementation tests   46–7,
      47, **48**, 257
concentration-gradient model
      31
connective tissue   209, 257
contents summaries [of
      literature]   4
corpus layer [of shoot]   *36*, 181
correlations between organs
      32–3
correlative enhancement   33
correlative inhibition   33

cosmids 57, 257
    effect on mutant phenotypes 58
cowpea mosaic virus, cell-to-cell movement 229, **230**
critical day length 194
crossing-over process 10
cross-protection 232–5, 257
crown galls 20
Cruciferae
    self-incompatibility 214
    see also Arabidopsis; Brassica
cryptochromes 128
cryptophyte algae 8–9
cucumber mosaic virus
    genome structure 234
    interaction between helper virus and satellite RNAs **233**
Cucumis melo [melon], genome size 102
Cucumis sativus [cucumber], sex change in flowers 136
cytokinins
    chemical structure 135
    effects 135, 136, 238

dark germinators 153
databases 4
day length
    critical 194
    measurement of 130, 194–5
day-neutral plants 194
defence reactions 221, **241**, 248
dehiscence 211
deletion analysis **78**, 79
desmotubulus 30
dessication phase [of seed development] 147–50
determination of cells 26–8
    meaning of term 25, 257
    rolling-ball model 26–7
development
    key processes 13–20
        differentiation 14, 17
        growth 14, 15–17
        morphogenesis 15, 18–20
    meaning of term 7, 9
    regulation in multicellular organisms 25–34
developmental competence 25, 28
developmental genes
    autonomous action 50–1

effects affected by gene expression 63, **64**
isolation of
    by chromosome walking 54–7
    by isolation of molecularly marked genes 58–9
    by map-based cloning 52
    by molecular identification of the gene 62
    via mutagenesis by T-DNA integration 59–60
    by RFLP mapping 54
    by tracking down the gene 57–8
    via transposon mutagenesis 60–1
    meaning of term 257
    molecular analyses of effects 62–4
    mutant phenotypes used to identify 42–52
    and regeneration 51
    time and place of effect 49–50
dexamethasone, trichome development affected by 189–90
differential gene expression 21–3
    plastid genes 101, 101
differentiation
    meaning of term 14, 15, 17, 257
    see also cell differentiation
diffusion reaction model 31
diploid organisms 10, 11
diploid phase 11
diplonts 11
distance, genetic 52, 257
DNA methylation 86
DNA polymorphisms 52, **53**, 257
DNA–protein complexes 7, 8
DNA re-association kinetics 71, 72, 257
dormancy, see also seed dormancy
dorsoventral symmetry 18
dot blot technique 115, **116**
double mutant analysis 172
double mutants, phenotypic analysis of **48**, 51, 51

down-regulation [of transcription] 120
Drosophila melanogaster [fruitfly] 1
    cascade of regulatory genes 83, 208
    genetic tumours 19
    genome size 70, 71

ectopic expression [of genes] **64**, 257
effect spectra [action spectra] 119, 128, 257–8
electrical signals, cell-to-cell communication via 32
electrical stimuli, cell differentiation affected by 20
electromagnetic radiation, as mutagen 44, 46
elongation rates [cell enlargement] 16
elongation zone [in root] 16, 176–7, 262
embryogenesis
    animal cells 32
    plant cells 13, 14, 163–7
        pattern formation during early embryogenesis 167–73
embryogeny 18–19
embryonic development, animals compared with plants 34–5
EMS (ethyl methane sulphonate), as mutagen 44, 58, 62
endonucleases 95
endoplasmatic reticulum 8, 30
endosymbiosis 8–9
endothelium [of ovule] 212, 257
enhancers 76, 85, 257
enlargement of cells 16–17, 29
ENOD40 [phytohormone] 135, 139
environmental factors
    effects on development phases 3, 111
    see also light; temperature
epidermis, root 174, 178, 179
Epifragus virginiana, plastome 90, 90, 97
epistasy **48**, 51, 257

Escherichia coli
  in gene transfer techniques
    243
  genome size *71*
ethylene
  as phytohormone *135*
  effects *135*, 141–4
ethylene signal transduction
    chain 142–4
etiolated plants
  greening of 98, 113–14, *114*
    changes in protein and
        mRNA levels *115*
    experiment techniques
        used to study **116**
    plastid differential study
        89, 115
eubacteria, cellular composition
    7–8
eukaryotes 7
  cellular composition 8
Eupatorium chinense, virus-
    infected 223
evocation 191, 257
exine [of pollen grain] 210,
    258
exonucleases 95

fasciation 185, 257
fate map *183*
female gametophyte, formation
  and development of
    211, 212–13
ferns, reproduction 12
fertilization *14*, 30, 163,
    214–15
flavonoids
  in bacterium–plant
    interactions 221,
    241, 248, *250*
  chemical structures *250*
florigen ['flowering hormone']
    194
flower formation 13, *14*,
    191–2
  factors affecting 193–6
  typical features *210*
flower-formation transcription
  factors *81*, 204–5
flowering-time genes 195–6,
    207
flower meristem 197, 198
flower meristem identity genes
    207, 258

flower organ identity genes
  expression of 205
  listed *203*
flower organs, production of
    14, 197–209
flower tube [in *Antirrhinum*]
    198, *199*
footprinting method **78**, *79*,
    258
formative growth 16
Fritillaria assyrica, genome size
    70
Fucus [alga], polarity of zygotes
    24, *25*
funiculus 163, *210*, 212, 258

galls 20
gametophytes 11, 258
  formation and development
    of 11, 13, *14*, 208–13
    female gametophyte
        212–13
    male gametophyte 209–11
gametophytic self-
    incompatibility 213
gel retardation method **78**, *79*
gene activity 23
gene expression
  differential 21–3
    plastid genes 101, *101*
  ectopic **64**
  flower organ identity genes
    205
  mitochondrial genes, factors
    affecting 103–4
  nuclear genes, factors
    affecting 84–6, 105–7
  phytochromes affecting
    119–20
  phytohormones affecting
    145
  plastid genes, factors affecting
    95–100, 105, *106*
  regulation of 63–4, 69
    by biological clock 131–3
    by blue light 129–30
    by phytochromes 119–20
    by phytohormones 142–5
gene families 73, 76
  phytochrome genes 121–2,
    122
generations
  alternation of 9–12
  meaning of term 11, 258

generative phase 191–215
  see also fertilization; flower
    formation;
    gametophytic phase;
    pollination; zygote
    formation
generative reproduction 10,
    12, 13
genes
  cell differentiation affected by
    20–1
  locating of 57–61
    via insertion mutagenesis
      59–60
    via transposon mutagenesis
      60–1
  molecular identification of
    62
  number in plant genome
    40, **41**
  see also developmental genes;
    differential gene
    expression;
    housekeeping genes
gene tagging 46, 52, 58, 258
genetic matter
  division into subgenomes 2,
    67–107
  see also chondriome; nuclear
    genome; plastome
genetic mosaics 50, 258
genetic terminology **48**
genetic tumours 19
gene-transfer mechanisms,
  agrobacterium–plant
    transfer *241*, 242–3
genome colinearity 74, *75*
genome organization 76–7
genome size, listed for various
    subgenomes/species *71,
    90, 102*
genome synteny 74, 262
giberellins
  chemical structure *135*
  effects *135*, 136, 144
Ginkgo biloba [maidenhair tree],
    plastome *90*
globulins **148**
glossary 256–62
Glycine max [soybean]
  rhizobium–legume symbiosis
    135, 139, 247, 251–2,
    252
  as short-day plant *194*

glycoproteins   24, 29
Gossypium [cotton], seed
          development   146–50
Gramineae [grasses]
     genome comparison   74, *75*
     photomorphogenesis
          113–14, *114*
     plastid differentiation   89, **89**
     seed development   146, *147*,
          148
     *see also Avena; bamboo;
          Hordeum; Oryza; Secale;
          Setaria; Sorghum;
          Triticum; Zea*
growth
     by cell division   15–16
     by cell enlargement   16–17,
          29
     meaning of term   14, *15*
growth factors   31–2
GT-1 transcription factor   *81*,
     120
guard cells, formation of   29,
     29

haploid phase   *11*
haplonts   11, 12
Hedera helix [ivy], cell
          differentiation   13, 27
helper viruses   232, **233**, 258
heterochrony genes   195, 207
high irradiance reaction   117,
     118
homeobox   258
homeobox proteins   207
homeodomain-containing
          proteins, transcription
          factors for   *81*
homeosis   **200**
homeostasis   258
homeotic genes   **200**, 258
homeotic mutants   199–200,
     200
     ABC model to explain
          phenotypes   199–2,
          202
Homo sapiens, genome size   71
Hordeum vulgare [barley]
     genome size   71
     photoperiodic control of
          flowering   194
     plastid differentiation in
          88–9
     plastid genes   98–9, *100*

hormones   3, 33, 133
     see also phytohormones
housekeeping genes   23, 40
hybrid bleaching   105, 258
Hyoscyamus niger [henbane]   194
hypermethylation   86
hypersensitive reaction   222,
     235, **236**, 258
hypomethylation   86

idioblast   17, 258
immunofluorescence/immuno-
          staining   **64**, 258
incrustation [of trichomes]   190
indirect immunofluorescence/
          immunostaining   63,
          64
indole-3-acetic acid   *135*
     synthesis affected by
          agrobacteria   238–9
induction reactions   117
     low-fluence reaction
          117–18
     very low fluence reaction
          117, 118, 119
infection-caused tumours
          19–20
inflorescence   **192**, 258
inflorescence meristem   192,
          192, 207
insertional mutagenesis   46,
          52, 58, 259
in situ/in vitro hybridization
          63, **64**, 204–5, 259
interkinesis   11
interphase   15–16
intine [of pollen grain]   210,
          259
inverse PCR technique   58, 259
in vitro fertilization   215
in vitro transcription activity,
          measurement of   **98**

jasmonic acid
     chemical structure   *135*
     effects   135, 137, *138*

Kalanchoe blossfeldiana   194
Kozak's rule   226, 259
     strategies to work around
          limitations   226–9

Lactuca sativa [lettuce]   194
lamina [blade of leaf]   186

lateral inhibition   190
leaf development   186–7
lectins   251, 259
leghaemoglobin   253, 259
legumes
     in rhizobium–legume
          symbiosis   244–53
     *see also Glycine; Medicago;
          Phaseolus; Pisum;
          Trifolium; Vicia*
leucine zipper   *81*, 259
life, basic attributes   9
life cycles
     and development   9–12
     higher land plants   12–13,
          14, 162
     simple   12
light
     characteristics   111–12
     developmental processes
          affected by   *112*
     as external factor   3, 111
     flowering affected by   *112*,
          193–5
     seed germination affected by
          112, 119, 153–4
     units used   112
light germinators   153–4
light-regulated [gene]
          promoters   120, *121*
light-regulation transcription
          factors   *81*, 120
light signal transduction chain
     cell biology approach
          123–4, *125*
     genetic approach   124–7
     'light switches'   120, *121*, 130
lilies, genome size   70
locus control region   85, 259
long-day plants   194
     examples   *194*
long-distance communication
          [between cells]   32–4
Lycopersicon esculentum [tomato]
     genome   71, 73
     phytohormone   135, 139

macrospores, *see* megaspores
MADS box   196, 203
MADS box proteins   *81*, 204,
          204
male gametophyte, formation
          and development of
          209–11, *211*

malignant tumours 19
map-based cloning 52, 259
MAP kinase cascade 142
Marchantia polymorpha
[liverwort], genomes
90, 101, *101*, *103*
markers 47, 52, **53**, 242–3,
259
MARs (matrix attachment
regions) 84
maternal transmission 88, 211
mechanical stimuli, cell
differentiation affected by
20
Medicago sativa [lucerne/alfalfa],
root nodules *244*, 247
megaspores *14*, 210, 212, 259
meiosis 10–11, 13, *210*
meiospores 11, 208
Mendelian rules, inheritance
not explained by 87–8
meristematic cells 35
differentiation of 17, *29*
see also root ...; shoot
meristem
metameric symmetry 18
metamorphosis, abnormal **200**
metaphase 16
Micrasterias alga, cell growth
16
micropyle 212, 259
microspores *14*, 210, *210*, 259
Mirabilis jalapa, non-Mendelian
inheritance 87–8
mitochondria 8
genomes, *see* chondriomes
mitochondrial genes, factors
affecting expression
103–4
mitosis 10, *210*
model systems 1, 40–1, 221
molecular markers 52, **53**
see also DNA polymorphisms
Morgan [genetic distance unit],
equivalence in physical
units 52
morphogenesis 18–20
effect of blue light 127
meaning of term 15, *15*, 18,
259
pathological 19–20
see also photomorphogenesis;
scotomorphogenesis
mosses, reproduction 12

movement proteins 229, **230**,
259
multipartite genomes 227–8,
228
mutagenesis
choice of mutagen 44, 46
with endogenous transposons
60
with heterologous
transposons 60–1, *61*
pollen mutagenesis 44, *45*
seed mutagenesis 42–4
see also insertion ...;
saturation ...;
transposonmutagenesis
mutagens 44, 46, 259
mutant alleles **48**
mutant phenotypes 40
developmental genes
recognized by 42–52
mutants, genetic
characterization of
46–7, 124–7
Myb proteins *81*, 121

NDH complex, genes affecting
100
negative-sense RNA 226
NEP (nuclear-encoded [RNA]
polymcrase) promoters
96, 97
Neurospora crassa fungus,
circadian-rhythm genes
132, *133*
Nicotiana glutinosa, virus
infection 235
Nicotiana sylvestris, virus
infection 235
Nicotiana tabacum [tobacco]
DNA/RNA hybridization
experiments 22
genome size 71, *90*
life cycle *14*
plastome *90*, *91*, *92*
transcription factors *81*
virus infection 235, *235*
see also tobacco mosaic virus
nod factors 139, 222, 248
chemical structures *251*
nod genes 247, 259
funcitons **251**
identification of **248**
nodulin genes 247, 252, 259
nopaline 237, *239*, 259

nopaline synthetase [enzyme]
239
Northern blot technique 94
Northern hybridization 62, 63
nucellus 212
nuclear genes
factors affecting expression
84–6, 106–7
structure 76–7
transcriptional control region
77–80
nuclear genomes 70–87
meaning of term 70
molecular anatomy 73–4
size and complexity 70–3
size listed for various
species *71*
nucleoid 7–8
nucleoides 88
nucleomorph 9
nucleus 8

octadecanoids
chemical structure *135*
effects *135*, 137
octant [pro-embryo] *164*, *165*
octopine 237, *239*, 259
octopine synthetase [enzyme]
239
Oenethera [evening primrose],
genome–plastome
interactions 105, *106*
oligogalacturonides *135*, 138
oligosaccharides
chemcial structure *135*
effects *135*, 137–9, 248, 249
onc genes 239, 242, 259
ontogeny 9–13
opaque 2 protein *81*, **148**
opines 237
chemical structure *239*
Oryza sativa [rice]
genome
colinearity 74, *75*
size *71*, *90*
transcription factors *81*
ovule 163, *210*, 213, 260

pararetrovirus 227
paternal transmission 88
pathogenesis-related proteins
221, 235
pathological morphogenesis
19–20

pattern
  definition   25, 260
  see also apical–basal pattern;
      branching; phyllotaxis;
      radial pattern
pattern formation   25, *26*,
      28–9, 167–73
PCR (polymerase chain
      reaciton) amplification
      53, 260
  see also inverse PCR; RT–PCR
pea, *see Pisum sativum*
Pelargonium
  genome size   *90*
  non-Mendelian inheritance
      88
PEP (plastid-encoded [RNA]
      polymerase) promoters
      96–7, *96*
peptide growth factors [in
      animal cells]   32, 260
peptide hormones   *136*
peribacteroid membrane   246,
      253
pericycle cells [in root]   *174*,
      177, 179, 260
Perilla ocymoides [*P. frutescens*]
      194
petiole [stalk of leaf]   186, 260
Petroselinum sativum [parsley],
      chalcone synthetase
      [gene] promoter   120,
      121
Phaseolus spp. [bean], root
      nodules   247
phlobaphenes, synthesis in *Zea
      mays*   82, 83
phosphoenolpyruvate
      carboxylase genes   77,
      107
photoaffinity labelling   140–1,
      260
photoinhibition   156, 260
photolyases   128
photomorphogenesis   19, 113,
      114, 124–5, 260
photon fluence   112
photon fluence rate   112
photoperiodicity   *112*, 193–5
photoreceptors   111, 112–14
photosynthesis, genes affecting
      100, *106*
photosynthesis apparatus,
      biogenesis of   155–7

photosystem II, components
      95, *95*, *106*, 131, 156
phyllotaxis   28, 31, 180, 260
phytochrome-dependent
      reactions   116–19
  high irradiance reaction
      117, 118
  induction reaction   117
phytochromes   114–119
  gene expression affected by
      119–20
  physiological functions
      122–3
  as products of gene family
      121–2
  signal-transduction chain
      123–4, *125*
  structure   115, *118*
phytohormone receptors   140–1
phytohormones   3, 33, 133–45
  antagonistic effects   136,
      138, 140
  binding proteins   141
  chemical structures   *135*
  effects   134–40
  as signal transporters   111,
      133
phytomer(e)s   35, 260
Pinus, plastome   *90*, 93
pistil   *210*, 211
Pisum sativum [garden pea]
  auxin regulator signals
      144–5
  genome size   *71*, *90*
  lectin gene   251
  root nodules   *244*, 247
  plasmid rescue   58, 260
plasmodesmata   *8*, 30, 32, 162
  virus movement via   229,
      229, **230**
plasticity, developmental   35
plastid differentiation   *89*
  study of   *89*, **89**
plastid DNAs   89–92
plastid gene promoters   95–7
plastid genes
  factors affecting expression
      95–100, 105, *106*
  organization in polycistronic
      transcription units
      93–4
plastid RNAs, relative stability
      99–100
plastids   8

differentiation of   17
genomes, *see* plastomes
  polyploid/polyenergid nature
      88–9
  as semi-autonomous
      organelles   93
plastomes   87–100
  nature of   88–9
  non-Mendelian inheritance
      87–8
  repertoire of genes in   92–3
polar coordinate model   31
polarity [of cell development]
      23–4, *25*, 33–4, 260
pollen mutagenesis   44, *45*
pollen sac   209, *210*
pollen–stigma interactions   30,
      211, 213–14
pollen tube   210, *210*, 214
pollination   *14*, 213–14
polycistronic transcription units
  plastid genes organized in
      93–4
  in tobacco genome   *91*
  transcription patterns   94–5
polyproteins, viral RNAs
      transcribed into   228,
      228, 260
positional information of cells
      30–2
  concentration-gradient model
      31
  diffusion reaction model   31
  polar coordinate model   31,
      31
post-embryonic [vegetative]
      development   176–91
post-transcriptional control
      mechanisms   69, 86–7
potato, *see Solanum tuberosum*
potato spindle tuber viroid,
      genome structure   *227*
primary leaves   180, 183, 260
primary root meristem, *see* root
      meristem
primary shoot meristem, *see*
      shoot meristem
prokaryotes   8
prolamellar bodies   **89**
prolamins   **148**
promoter, meaning of term
      76, 260
prophase   16
prothallus   12

protoderm 166, 260
Prototheca wickerhamii [green
        alga], genome size *101*
proximo-distal axis 186, 260
PR (pathogenesis-related)
        proteins 221, 235, 260
psbB operon *91*, 94–5, *94*
psbD/C operon 95, *96*
publications 3–4
    lists *5*

radial pattern [in seedlings]
        167
    mutants in *Arabidopsis*
        171–3
radial symmetry 18
RAPDs (randomly amplified
        polymorphic DNAs) **53**,
        260
rbcS gene family 73, 76
receptors
    photoreceptors 111, 112–14
    phytohormone–receptor
        complexes 140
recognition reactions 221
re-embryonalization 17
reporter genes **64**, 261
reproduction
    asexual/vegetative 10, 12,
        13
    sexual/generative 10, 12, 13
research developments 2
review articles/publications 4,
        5
RFLP (restriciton fragment-
        length polymorphism)
        technique **53**, 54, 261
    RFLP mapping 54, *55*, 74
    RFLP markers 54, 74
rhizobia 244–53
    see also Azorhizobium;
        Bradyrhizobium;
        Rhizobium
rhizobium–legume symbiosis
        3, 30, 244–53
    oligosaccharides 138–9, 248
    requirements for
        establishment 252–3
Rhizobium spp.
    effect of flavonoids *250*
    genetic maps of plasmids
        249
    nod factors, chemical
        structures *251*

ribosomal proteins, genes
        affecting *100*
RNA editing 103–4
RNA polymerase
    genes affecting *100*
    mitochondrial 103
    nuclear- compared with
        plastid-encoded 97–8
RNA virus genomes,
        positive/negative
        designation 226
rolling-ball model 26–7
root
    adventitious 180
    cell enlargement during
        growth 16, 176
    embryonic 175
    lateral 35, 179–80
    primary 176–7
root hair cell 178, 261
    morphogenesis 179, *179*
root hairs, development of
        178–9
root meristem *35*, 167, 261
    organization 174–5, *174*,
        176
    origin 173–4
root nodules 244, *244*
    development of *245*, 246–7
    structure 245–6, *245*
RT-PCR (reverse
        transcription–polymerase
        chain reaction) technique
        212, 257
Rubisco (ribulose-1, 5-
        biphosphate
        carboxylase/oxygenase),
        genes coding for 73, 76
Rumex [sorrel] 214
run-on transcription [rate
        measurement] method
        98, *99*, 119

Saccharomyces cerevisiae [baker's
        yeast]
    genome size *71*, 101
    osmoregulation *143*
Salix [willow], cell polarity
        33–4
SARs (scaffold attachment
        regions) 84–5, *85*
satellite RNAs 232, **233**, *234*,
        261
satellite viruses 232

saturation mutagenesis 42, 46,
        48, 261
scotomorphogenesis 19, 113,
        114, 124, 126, 260
SDS–polyacrylamide gel
        electrophoresis *115*,
        116
Secale cereale [rye], shoot-
        elongation rates 16
seed, definiton 145
seed development
    abscisic acid affecting 151–2
    global regulators 149–51
    phytohormones and light as
        regulators 145–57
seed dormancy 149, 152–3
    breaking of 153
    ecological advantage 153
seed germination 13, *14*,
        152–3
    induction by light *119*, 153
seedlings
    body organization 167–73
    development of 153–7
seed mutagenesis 42–4
seed production 145–50
seed storage proteins **148**
segregational analysis 59, 261
self-incompatibility mechanisms
        213
sequence deletions **78**, *79*
Setaria viridis [millet] *194*
sex-determiantion mechanisms
        214
sexual reproduction 10
shade avoidance syndrome
        112, 123
shaded plants
    effect of light on development
        112
    role of phytochrome system
        117
    see also etiolated plants
shape changes, virus-caused
        224
shikimate pathway, enzyme
        73
shoot development 16, 180–1
shoot merisem *35*, 162, *167*
    organization 36, 181–6, *181*
    origin 175–6
short-day plants 194
    examples *194*
sieve tubes 32

signal molecules   111, 241
see also phytochromes;
   phytohormones
signal transduction chains
   phytochromes   123–4, *125*
   phytohormone–receptor
      interactions   140–4
   rhizobium–legume symbiosis
      248, *249*
   virus–host interacitons   221,
      236
Silene [campion]   214
Sinapis alba [white mustard],
   phytochromes   130
Solanaceae
   self-incompatibility   213
   see also *Hyoscyamus*; *Nicotiana*;
      Solanum
Solanum tuberosum [potato],
   genome size   71
Sorghum bicolor
   differential expression of
      plastid genes   **100**,
      100
   genome size   71
Southern hybridization   **53**,
   74
soybean, *see* *Glycine max*
Spinacia oleracea, plastid DNA
   90
spores   10, *14*, 210, *210*, 212,
   261
sporoderm   210, 261
sporophytic   phase   12–13,
   14, 213, 261
sporophytic self-incompatibility
   213
sporpollenins   210, 261
SSLPs (simple sequence-length
   polymorphisms)   **53**
stem cells   35, 261
stomatal development   29, *29*
storage carbohydrates   146
   degradation and mobilization
      of   154
storage lipids, degradation and
   mobilization of   154
storage proteins   **148**
   degradation and mobilization
      of   154
stretching of cells   16
subgenomes
   interactions between   104–7

structure and expression   2,
   67–104
   chondriome   101–4
   nuclear genome   70–87
   plastome   87–100
   see also chondriome; nuclear
      genome; plastome
subgenomic RNA, viral RNA
   227, *228*
suspensor cells   163, 164, *165*,
   174, 261
switches
   light-regulated   120, *121*,
      130
   vegetative-to-generative
      191
symbiosis, *see*
   rhizobium–legume
   symbiosis
symmetries, ontogenetic   18
symplast   162
symptoms
   meaning of term   223
   virus-caused   224–5
syngamy   10, *11*
synteny   74, 261
systemine   *135*, 139

TATA box   76, *80*, *85*, 261
telophase   16
temperature
   as external factor   111,
      130–1
   flower formation affected by
      193
   seed germination affected by
      153
temperature-sensitive alleles
   49–50
thecae   209, 261
theoretical resolution [between
   gene and marker]   56,
   261
Ti (tumour-inducing) plasmid
   237, *238*
   derivatives as gene vectors
      242
   genome structure   *238*
   T-DNA   237, *238*, *239*
      mobilization of   *242*
   vir region   237, *238*, *240*
      functions   *240*, **241**
tobacco, *see* *Nicotiana tabacum*

tobacco mosaic virus
   cell-to-cell movement   229,
      230
   genome structure   227
tomato, *see* *Lycopersicon
   esculentum*
tomato spotted wilt virus,
   genome structure   227
totipotency   17
tracheary elements   17, ??
Tradescantia, development of
   stomata   29
transcriptional control regions,
   identification of   **78**
transcription factors   76
   interaction in *Zea mays*   81–4
   listed for various species   *81*
   in rhizobium–legume
      symbiosis   248, *249*
transcription rates
   plastid genes
      in barley seedlings   *100*
      measurement by run-on
         experiments   **98**, *99*
transcription start points   **78**, *79*
transient expression systems
   78
translational frameshift   *228*,
   229
transposon mutagenesis   60–1,
   197, **236**, **248**
transposons   60, 261
trans-regulatory factors
   identification of binding sites
      78, *79*
   isolation from cDNA
      expression libraries   **78**
trichomes
   development of   50, 187–90
   mutant phenotypes   189–90,
      191
Trifolium pratense [clover],
   susceptibility to rhizobial
   infection   251
triple response   141
Triticum aestivum [wheat]
   gene families   76
   genome   *71*, 73
Tulipa spp.
   genome size   71
   virus-infected   223, *223*
tumours   19–20
   induction by agrobacteria
      237–42

tunica layer [of shoot]   *36*,
    181, 262

UV (ultraviolet) radiation
    receptors   127
    UV-A/UV-B regions   112, 127

vegetative development
    176–91
vegetative reproduction   10,
    12, 13
vernalization   153, 193, 262
Vicia faba [broad bean]
    as long-day plant   *194*
    root nodules   247
vir genes   241–2, 262
    functions   *240*, **241**
viroids   222, 232
    compared with viruses   225,
    227
vir region [of Ti plasmid]   237,
    238
viruses   222–35
    cell-to-cell movement   229,
    230

cross-protection by   232–5,
    257
genomes   225–9
infection first described   223
nomenclature   224
resistance acquisition   235
structures   225, *226*
symptoms   224
    explanations   230–2
Volvox alga, cell polarity   33
VP1 (viviparous 1) protein   *81*,
    149

wound-induced signal
    compounds   **241**, 248
    see also flavonoids

Xanthium strumarium
    [cocklebur]   194, *194*

YAC (yeast artificial
    chromosome)   **41**, 57,
    262
    clones   *56*, 57
yeasts, *see Saccharomyces cerevisiae*

Zea mays [maize]
    chloroplast biogenesis   156
    genome
        analysis   70, *72*
    comparison with other
        grasses   74, *75*
    size   *71*
    genome size   *102*
    as model system   1
    sex-determination
        mechanism   214
    transcription factors   *81*
    transposons   60
zeatin   *135*
zinc finger proteins,
        transcription factors for
        81
Zinnia spp., cell differentation in
        17, 22
zone of elongation [in root]
        16, 176–7, 262
zygomorphic flower   198, 262
zygote formation   10, *14*, 163,
        214